LANDSCAPE PATTERN ANALYSIS FOR ASSESSING ECOSYSTEM CONDITION

Environmental and Ecological Statistics

Series Editor
G. P. Patil

Center for Statistical Ecology and Environmental Statistics
Department of Statistics
The Pennsylvania State University
University Park, PA, 16802, USA
Email address: gpp@stat.psu.edu

The Springer series **Environmental and Ecological Statistics** is devoted to the cross-disciplinary subject area of environmental and ecological statistics discussing important topics and themes in statistical ecology, environmental statistics, and relevant risk analysis. Emphasis is focused on applied mathematical statistics, statistical methodology, data interpretation and improvement for future use, with a view to advance statistics for environment, ecology, and environmental health, and to advance environmental theory and practice using valid statistics.

Each volume in the **Environmental and Ecological Statistics** series is based on the appropriateness of the statistical methodology to the particular environmental and ecological problem area, within the context of contemporary environmental issues and the associated statistical tools, concepts, and methods.

Additional information about this series can be obtained from our website:
www.springer.com

LANDSCAPE PATTERN ANALYSIS FOR ASSESSING ECOSYSTEM CONDITION

by

Glen D. Johnson
New York State Department of Health and
University at Albany, State University of New York, USA

and

Ganapati P. Patil
Center for Statistical Ecology and Environmental Statistics
The Pennsylvania State University, USA

 Springer

Library of Congress Control Number: 2006930744

ISBN-10: 0-387-37684-4   e-ISBN-10: 0-387-37685-2
ISBN-13: 978-0-387-37684-4

Printed on acid-free paper.

Printed in the United States of America.

9 8 7 6 5 4 3 2 1

springer.com

Contents

List of Figures

List of Tables

Preface

Consider an imminent 21st century scenario: What message does a multi-categorical map have about the large landscape it represents? And at what scale, and at what level of detail? Does the spatial pattern of the map reveal any societal, ecological, environmental condition of the landscape? And therefore can it be an indicator of change? How do you automate the assessment of spatial structure and behavior of change to discover critical areas, hot spots, and their corridors? Is the map accurate? How accurate is it? How do you assess the accuracy of the map? How do we evaluate a temporal change map for change detection? What are the implications of the kind and amount of change and accuracy on what matters, whether climate change, carbon emission, water resources, urban sprawl, biodiversity, indicator species, human health, or early warning? And with what confidence? The involved research initiatives are expected to find answers to these questions and a few more that involve multi-categorical raster maps based on remote sensing and other geospatial data.

One of our greatest challenges as we begin the 21st century is the preservation and remediation of ecosystem integrity. This requires monitoring and assessment over large geographic areas, repeatedly over time, and therefore cannot be practically fulfilled by field measurements alone. The vastly under-utilized images from remote sensing may therefore prove crucial by their ability to monitor large spatially continuous areas. This technology increasingly provides extensive spatial-temporal data; however, the challenge is to extract meaningful environmental information from such extensive data.

In this monograph, we present a new method for assessing spatial pattern in raster land cover maps based on satellite imagery in a way that incorporates multiple pixel resolutions. This is combined with more conventional single-resolution measurements of spatial pattern and simple non-spatial land cover proportions to assess predictability of both surface water quality and ecological integrity within watersheds of the state of Pennsylvania (USA). The efficiency of remote sensing for rapidly assessing large areas is realized through the ability to explain much of the variability of field observations that took several years and many people to obtain.

GLEN D. JOHNSON AND GANAPATI P. PATIL

Acknowledgments

Over the years, this monograph and the related research have been supported directly or indirectly by the grants from the National Science Foundation and the Environmental Protection Agency. The NSF grants have been awarded by the Statistics and Probability Program, Environmental Biology Program and the Digital Government Program. The EPA grants have been awarded by the Office of Research and Development and by the Office of Policy, Planning, and Evaluation. The work has been carried out at the Center for Statistical Ecology and Environmental Statistics in the Department of Statistics at The Pennsylvania State University in collaboration with the Office of Remote Sensing of Earth Resources in the Penn State Institutes of the Environment. We wish to acknowledge the support and encouragement received from the program managers, Lawrence Brandt, Penelope Firth, and Sallie Keller-McNulty at NSF and the program managers, Barbara Levinson, Thomas Mace, Barry Nussbaum, and Phillip Ross at EPA. And at Penn State, James Rosenberger, Archie McDonnell, and William Easterling.

Over the years, we have had stimulating interactions with our colleagues Wayne Myers and Charles Taillie. We greatly appreciate and recognize the valuable long time collaboration. The material has been used in one form or the other in the cross-disciplinary ecometrics and environmetrics classroom, while addressing issues and approaches involving landscape pattern analysis for assessing ecosystem condition.

Over the years, we have been extremely fortunate to have had Barbara Freed with us to assist in a manner that has been only delightful. Our heartfelt appreciation for her perceptive devotion.

Finally, we acknowledge our families and friends for their support and encouragement in the writing of this book over these years.

Glen D. Johnson
Ganapati P. Patil

Chapter 1

INTRODUCTION

Assessment and monitoring of ecosystems at the landscape scale requires characterization of spatial land cover patterns. Ideally, we want a quantitative basis for deciding when a landscape pattern undergoes substantial changes over space and/or time.

Of key importance is identifying ecosystems that are close to a critical point of transition into a different, possibly degraded, condition. For a native ecosystem of forest, degradation could mean that the landscape matrix has become developed non-forest land, supporting only small sparsely scattered forest islands that do not provide sufficient forest interior habitat. Along with the collapse of forest-interior species richness, degradation may also be evidenced by increasing environmental contamination that is also associated with intensive land development.

Landscape fragmentation can be a primary component of an *ecosystem risk assessment* in a similar manner as addressed by Graham, Hunsaker, O'Neill and Jackson (1991) and Carlsen, Coty and Kercher (2004). Identifying landscape areas where ecosystems are poised for a great reduction in overall species and/or the elimination of critical functional groups is of particular concern because such areas may still be salvaged with intervention by proper land use planning. Meanwhile, other areas that have "crossed the line" but are not too far beyond the critical degradation point may still be reversible. Indeed, ecosystems that are near critical transition points present both risks and opportunities.

The effects of landscape-scale land cover/land use pattern on both biodiversity and water quality are now further elaborated, followed by a brief discussion of approaches to quantifying landscape pattern. Watershed-delineated landscapes in the state of Pennsylvania (USA) were used for application and evaluation of the methods discussed in this monograph.

Characteristics of Pennsylvania and the data used are introduced at the end of this chapter.

Chapter 2 presents methods for quantitatively characterizing landscape pattern in raster land cover maps. Along with conventional single-resolution measurements, Chapter 2 also presents an innovative method of multi-resolution conditional entropy profiles, which are further illustrated in Chapter 3. Chapters 4 to 6 evaluate the feasibility of using conditional entropy profiles and other measurements of landscape pattern for classifying landscape types and also for predicting ecosystem condition with respect to water quality (Chapter 5) and ecological integrity (Chapter 6). Finally, Chapter 7 summarizes key findings from throughout this monograph and identifies potential future directions.

1.   Loss of Biodiversity through Excessive Habitat Fragmentation

Conservation Biology is a scientific field that has grown from the rather grave concern about decreasing biodiversity, both locally and globally (Brooks and Kennedy, 2004). While the term "biodiversity" evades specific definition because of its broad usage throughout the years (Kaennel, 1998), what is clear is that animal species appear to be disappearing at rates never before experienced (Levin, Grenfell, Hastings and Perelson, 1997). Furthermore, while plant species appear to be somewhat more resistant, there is strong concern that plant extinctions are poised for a large increase in the near future (Gentry, 1996).

Such an observation raises the primary concern of reduced genetic diversity and ecosystem functional diversity (Frankham, 2005; DeSalle and Amato, 2004), both of which may occur at many spatial scales, ranging from small preserves of a few hectares to the entire planet. Actually the patterns of species composition and richness at some scales can influence patterns at other scales. For example, decline of both the number of different forest bird species and the population sizes within remaining species has been documented for some wildlife preserves where land use within the preserve is unchanged but conditions in the surrounding landscape matrix have become degraded (Askins, 1995).

Managing land use for maximizing overall species richness may seem to be an admirable objective. However, while the overall species richness may increase from practices that increase habitat diversity, this may not reflect the natural species richness potential under pristine conditions. For example, if an area falls within an ecoregion that is characterized by a potential natural vegetation of Appalachian oak forest, then continuous forest with some background patchiness from natural disturbance would represent natural habitat type under pristine conditions. As the forest is fragmented due to human activity, the "edge effect" may increase overall species richness; however, this may be to the detriment of native forest interior species due to increased predation and parasitism by opportunistic species of the forest edge (Brittingham and Temple, 1983; Noss, 1983; Yahner, 1988). Evidence is also cited by Yazvenko and Rapport (1996) that biodiversity may be higher in moderately disturbed sites compared to pristine sites, although severely damaged ecosystems seem to inevitably decrease in biodiversity.

Severe forest fragmentation results in *islands* of forest habitat in a *matrix* of open land types. Survival of forest interior species in such islands is possible if an island is large enough to maintain a sufficient buffer against the impact of opportunistic edge species, or if the island is sufficiently close to other "source" islands that are large enough to supply replacement individuals of a population (MacArthur and Wilson, 1967; Noss, 1983; Askins, 1995; Moquet and Loreau, 2003). In other words, a species may persist in a fragmented landscape as a meta-population distributed across habitat islands if the right conditions exist for that species (Hanski and Ovaskainen, 2000).

Conservation efforts are increasingly directed towards maintaining connectivity among islands of suitable habitat (Noss, 1996; Beier and Noss, 1998; Kaiser, J, 2001) in order to maintain reproducing populations of native species. Connectivity is itself scale-dependent, since the maximum distance that renders islands "connected" depends on the dispersal ability of the affected species. For example, Keitt, Urban and Milne (1997) quantified the connectivity of mixed-conifer and ponderosa pine habitat clusters in the southwest United States, concluding that for a species to perceive a cluster as connected, the species must be capable of dispersing more than 45 kilometers over inhospitable terrain. Other research (Pearson, Turner, Gardner and O'Neill, 1996) supports the observation that as the overall *proportion* of suitable habitat decreases, connectivity among suitable habitat islands becomes increasingly dependent on the *spatial pattern* of habitat distribution.

As can be seen from this discussion, for a given geographic region where the potential natural vegetation is continuous forest, overall species richness may increase as the forest becomes fragmented by human activity, but after a critical amount of forest habitat disappears, overall species richness can decrease, possibly to levels below what might be expected under undisturbed conditions. Furthermore, although overall species richness may be maximized by maximizing habitat diversity through a certain amount of forest fragmentation, the richness of certain key functional groups of organisms may be decreased. For example,

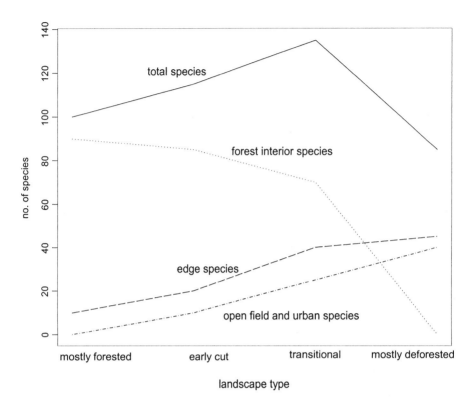

Figure 1.1. A hypothetical response of vertebrate species richness to increasing forest fragmentation.

Sekercioglu, *et al.* (2002) observed decreasing insectivorous birds from tropical forest patches.

A conceptual illustration of these ideas appears in Figure 1.1. Franklin and Forman (1987) speculate a similar scenario and further document other related forest impacts that can increase geometrically with increasing loss of forest. In Pennsylvania, Johnson, Myers, Patil and Walrath (1998) observed similar trends in regions defined by 635 km^2 EMAP hexagons, where the overall richness of breeding birds appeared to increase as the forest became increasingly fragmented, but a large relative drop in species richness was observed as the landscape became dominated by agricultural land.

Although the arguments thus far have been presented with respect to forested native ecosystems, they may also apply in other ecosystems such as prairie grasslands (Johnson, Igl and Dechant Shaffer, 2004) that are encroached by agriculture. Woody vegetation may even be invasive and problematic in a non-forest native ecosystem.

2. Land Use Pattern and Surface Water Quality

Ecological hierarchy theory proposes a framework for explaining how large-scale characteristics of ecosystems can constrain smaller-scaled characteristics (Urban, O'Neill and Shugart, 1987; O'Neill, Johnson and King, 1989). An example of such an inter-scale environmental relationship is the influence of gross land use characteristics on local surface water quality. Indeed, for all the improvements in water quality associated with modern controls on point-source discharges, local water quality is still constrained by non point-source pollution. Since land use is generally reflected by land cover (vegetation type), then whole watersheds may be evaluated with respect to water quality risk by characterizing land cover proportions and patterns (O'Neill, *et al.*, 1997; Mehaffey, *et al.*, 2003). Watershed-wide landscape characteristics that are significantly correlated with local water quality may then serve as landscape-scale indicators of environmental condition (Aspinall and Pearson, 2000; Jones, *et al.*, 1997).

Previous research suggests that for larger watersheds, corresponding to higher order streams, land cover proportions alone explain most of the water quality variability; whereas for smaller watersheds, especially those for first order headwater streams, the spatial *pattern* of land cover becomes more important (Graham, Hunsaker, O'Neill and Jackson, 1991; Hunsaker and Levine, 1995; Roth, Allan and Erickson, 1996). Thus, the feasibility of using watershed-wide marginal land cover proportions instead of more detailed spatial pattern characteristics for predicting water quality may depend on the hierarchical scale of the watershed.

3. Quantifying Landscape Pattern

Issues such as those discussed so far have motivated the development of many landscape measurements, as elaborated upon in Chapter 2. These measurements are quite valuable in their own right for obtaining information about specific landscape variables; however, one can rapidly get confused about what are the appropriate variables to be measuring when the objective is to categorize different landscapes into common groups. Such an objective may arise when many landscape units need to be assessed over a large region for the purpose of prioritizing units requiring intervention or remediation with respect to preventing or reversing environmental and ecological degradation (Jones, *et al.*, 1997).

Similarly, for any landscape unit that is being monitored over time, one may desire to categorize the status of the unit to determine if and how it may be changing. Therefore, one must decide what set of variables are appropriate to measure for achieving a desired ability to discriminate among different landscape types.

A further complication is to decide on an appropriate measurement scale (or grain) and an appropriate spatial extent of a landscape unit. Both of these aspects of spatial scale are known to affect observed patterns in landscapes (Wiens, 1989). In this monograph, we focus on landscapes that have been characterized by remotely sensed imagery, thus rendering a map of equal size and shape picture elements (pixels) that constitute a raster (or grid) type of map. Therefore, measurement scale is the resolution (size) of a data pixel, which is fixed by properties of a satellite sensor. Resolution is well known to affect observed pattern, as well as observed pattern change (Townshend and Justice, 1988; Qi and Wu, 1996).

Instead of letting resolution be a problem, an alternative approach is to employ the effects of variable resolution as being even more informative about spatial pattern than characterizations based on a fixed resolution. This approach motivated several researchers to investigate the application of fractal-based information theoretic measurements, as summarized by Johnson, Tempelman and Patil (1995), who also propose conditional entropy in this context. While detailed in Chapter 2, the basic idea is that multi-scale features of a land cover map may be captured in a sequence of successively coarsened raster map resolutions. The entropy of spatial pattern associated with a particular pixel resolution is calculated, conditional on the pattern of the next coarser "parent" resolution. When the entropy is plotted as a function of changing resolution, we obtain a simple two-dimensional graph called a "conditional entropy profile", thus providing a graphical visualization of multi-scale fragmentation patterns of a landscape within a fixed geographic extent (Johnson and Patil, 1998; Johnson, Myers, Patil and Taillie, 1999). This can be applied to landscapes defined by multiple land cover types and presents a more holistic, general assessment of pattern that is expected to encompass many aspects of pattern that would otherwise require a suite of different measurements.

The most closely related conventional measurement would be contagion; however, conditional entropy quantifies the spatial pattern at different resolutions in a way that captures the pattern's probabilistic dependence on the pattern of the next coarser resolution. This approach is more in line with hierarchy theory (O'Neill, Johnson and King, 1989) which suggests that patterns observed at a given resolution will constrain the patterns observed at finer resolutions and will in turn be constrained by coarser resolution patterns. Hierarchically-nested spatial patterns have indeed been observed in actual landscapes (Kotliar and Wiens, 1990; O'Neill, Gardner and Turner, 1992; Ernoult, Bureau and Poudevigne, 2003). Furthermore, Frohn (1998, pp. 69 and 75) showed that when contagion was measured over a range of resolutions that were obtained from an "increasing size majority filter", the results were an unpredictable trend in contagion as a function of increasing pixel size. This indicates that contagion may not be a viable candidate for obtaining a multi-resolution characterization of spatial pattern.

4. Application to Watershed-Delineated Landscapes in Pennsylvania

Conditional entropy profiles and single-resolution landscape measurements were obtained for 102 watersheds in the state of Pennsylvania (USA), which is portrayed in Figure 1.2. As elaborated in later chapters, results were used to cluster watersheds into common landscape types, then to test the general hypothesis that different indicators of ecosystem health respond to changes in landscape characteristics. A major motivation was to evaluate the feasibility of using remotely-sensed data as a primary means of assessing ecosystem condition.

Pennsylvania was chosen for initial application primarily because of the authors' ready access to extensive statewide databases and strong familiarity with this State. Furthermore, Pennsylvania presents a variety of different physiographic regimes that influence landscape-sculpting human activity in different ways.

The State of Pennsylvania can be conveniently stratified into three major physiographic provinces, as discussed by Miller (1995). The largest is the *Appalachian Plateaus*, which covers the western and northern part of Pennsylvania, encompassing about one half of the State's land area. This province is an upland area that is highly dissected by streams and rivers. Since the hilly topography discourages both agriculture and road building, human population densities are very uneven across the Appalachian Plateaus. Most of the province's Pennsylvania population lives within 100 miles of Pittsburgh, where development was encouraged by extensive bituminous coal deposits. The northern sector of the Appalachian Plateaus has no coal and its rough topography is not conducive to farming, therefore this region is sparsely populated. Although most of the northern sector has a history of logging, the current landscape matrix is largely mature forest land. These more pristine areas present a set of "background" watersheds against which others can be compared.

The *Ridge and Valley* physiographic province encompasses the second largest area of Pennsylvania, curving in a northeasterly direction from the southern border to the eastern border. The topography of this province is very distinctive due to its long, narrow ridges that are separated by equally long, but wide, valleys. The steep ridges have thin topsoil and are therefore not suitable for farming or most development; therefore, they remain mostly forested. Meanwhile, the largely limestone valleys have been cleared of forest to make way for farming and other development.

The smallest of the three major provinces is the *Piedmont Plateau* in the southeast corner of Pennsylvania. This is a gently rolling, well-drained plain that is seldom more than 150 meters above sea level. Given that some of the best soil in the eastern United States occurs here, this province is highly developed with farms, urban centers and increasingly more suburban sprawl. Further, the Piedmont encompasses Philadelphia, the nation's fifth largest city, and lies in the New York City-to-Washington D.C. corridor.

4.1   Land Cover Grids

The Pennsylvania land cover maps used in this monograph were derived from 6-band 30-meter resolution LANDSAT Thematic Mapper images. Information concerning the data can be obtained through the Pennsylvania Spatial Data Access web page (http://www.pasda.psu.edu, accessed January, 2006).

Each 30-meter pixel is assigned one of eight land cover categories, which are "water", "conifer forest", "mixed forest", "broadleaf forest", "transitional", "perennial herbaceous", "annual herbaceous" and "terrestrial unvegetated". The first four categories are self-explanatory. The transitional category results from a heterogeneous mix of land cover types; perennial herbaceous is primarily grassland and occurs in small patches throughout the state, but occurs in larger patches primarily where pastureland is present; annual herbaceous is generally cropland and is often adjacent to patches of perennial herbaceous land; terrestrial unvegetated is primarily urbanized land along with quarries and other rock rubble. Figure 1.3 is a land cover raster map of Pennsylvania, also showing the State Water Plan watersheds and the three major physiographic provinces.

The land cover classification methodology is an unconventional hybrid of supervised and unsupervised approaches. Six bands of thematic mapper imagery were first compressed into 255 clusters using the ERDAS IMAGINE© version of ISODATA with default seeds and single row/column steps. Labeling of clusters was done by supervised classification of cluster centroids using custom software that implements an adaptive Euclidean distance protocol. This approach has since been reconfigured in C-language programs for general use under the title "Pixel Hyperclusters As Segmented Environmental Signals", or PHASES (Myers, 1999).

4.2 Watersheds based on the State Water Management Plan

Watershed delineations were obtained from the Penn State Institutes of the Environment (PSIE), where small watersheds were aggregated to generally correspond to larger Pennsylvania State Water Plan watersheds. The source small watersheds were originally delineated by the Water Resources Division of the U.S. Geological Survey (USGS) on 7.5 minute topographic maps, then digitized to produce 9,895 digital drainage areas in Pennsylvania.

Although the original State Water Plan consists of 104 watersheds, the delineation obtained from PSIE consists of 102 watersheds. In the process of aggregating the smaller watersheds, two particular larger watersheds were created that each comprised the equivalent of two original state water plan watersheds. Nevertheless, the aggregated watersheds were preferred over the original state water plan delineation for two main reasons: first, the boundaries based on the aggregated small watersheds were much more spatially accurate; second, it was anticipated that future landscape analyses would be done within both smaller and larger watersheds that would be based on the original 9,895 small watersheds or some aggregation thereof, and it is desired to have smaller watersheds that are hierarchically nested within larger ones.

A coverage has since been made available (http://www.pasda.psu.edu) that consists of the original 104 watersheds that correspond to the state water plan, whereby the coverage was overlaid on the USGS-derived small watershed coverage and boundaries of the major watersheds were corrected to match. However, this corrected state water plan coverage was not yet available when this study was under way.

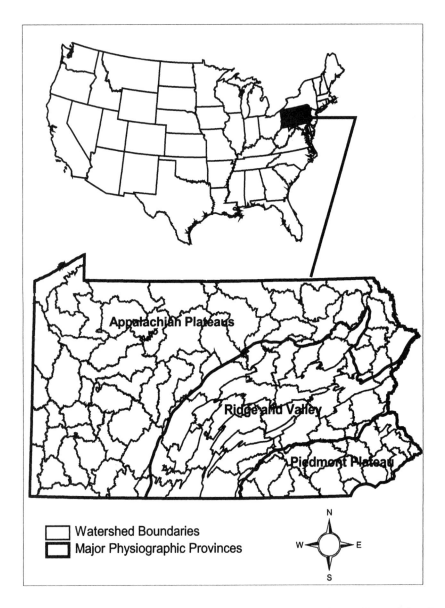

Figure 1.2. Location of the state of Pennsylvania within the United States (above) and watersheds based on the state water plan along with major physiographic provinces (below).

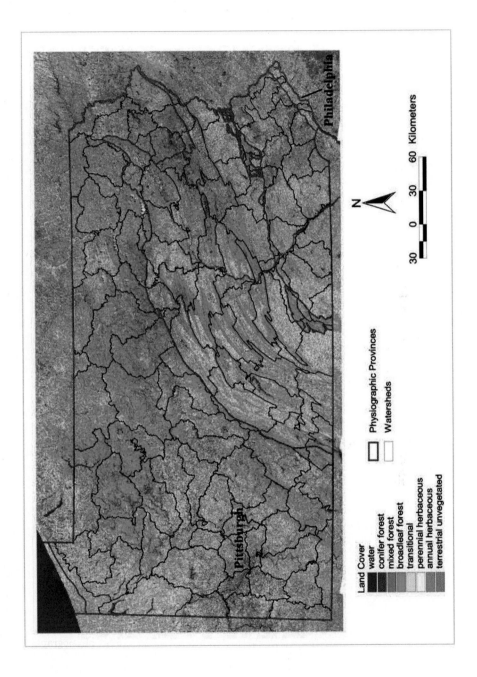

Figure 1.3. Land cover based on LANDSAT TM images, with watersheds based on the state water plan and physiographic provinces for Pennsylvania.

Chapter 2

METHODS FOR QUANTITATIVE CHARACTERIZATION OF LANDSCAPE PATTERN

1. Single-resolution Measurements

Once a land cover map is in hand, the simplest characterization of a landscape is the marginal (non-spatial) land cover distribution which is estimated from the readily available land cover proportions. Although the marginal distribution may be tightly linked to spatial pattern (Gustafson and Parker, 1992), two very different landscapes can still yield the same marginal distribution; therefore, additional measurements of specific aspects of spatial pattern are needed to properly characterize landscapes.

Many measurements of landscape pattern are available and most have been conveniently coded and made available as public domain software under the title of "FRAGSTATS" (McGarigal and Marks, 1995; www.umass.edu/landeco/research/fragstats/fragstats.html, accessed January, 2006). Many of these measurements are highly inter-correlated (Riitters, *et al.*, 1995; Hargis, Bissonette and David, 1998); therefore, one typically wants to select an appropriate subset.

The initial set of landscape variables that are applied in later chapters for the Pennsylvania watersheds are listed in Table 2.1. These measurements are based on eight different land cover types, as discussed in Section 4.1 of Chapter 1. Table 2.1 also includes land cover proportion summaries that were included with the pattern variables for subsequent analytical purposes. The land covers were summarized as follows: All forest cover types (conifer, mixed and deciduous) were summed to yield the proportion of "Total Forest" cover, and annual herbaceous and perennial herbaceous land was summed to yield "Total Herbaceous" cover.

Terrestrial unvegetated land was also included to capture much of the urban land.

Table 2.1.   Landscape variables measured for Pennsylvania watersheds.

Variable Description	Code
Patch Density	PD
Mean Patch Size	MPS
Patch Size Coefficient of Variation	PSCV
Edge Density	ED
Landscape Shape Index	LSI
Area-Weighted Mean Shape Index	AWMSI
Double-Log Fractal Dimension	DLFD
Area-Weighted Mean Patch Fractal Dimension	AWMPFD
Shannon Evenness Index	SHEI
Interspersion and Juxtaposition Index	IJI
Contagion[†]	CONTAG
Total Forest Cover	TOT.FOREST
Total Herbaceous Cover	TOT.HERB
Terrestrial Unvegetated	TU

note that diagonally adjacent pixels were included when determining patches
† pixel order preserved when measuring contagion

The spatial pattern measurements in Table 2.1 represent diverse aspects of pattern such as responses to average patch size, patch size distribution and patch shape complexity. *Patch density* and *edge density* are the number of distinct patches or edges, respectively, divided by the total landscape area. The *mean patch size* is the total landscape area divided by the number of distinct patches. The *patch size coefficient of variation* is the patch size standard deviation divided by the mean patch size.

The *landscape shape index* is the sum of the landscape boundary and all edge segments within the landscape boundary divided by the square root of the total landscape area, adjusted by a constant for a square standard. To formulate, let E equal the total length of edge, including the entire landscape boundary and background edge segments, regardless of whether or not they represent true edge, and let A equal the total landscape area. Then, for a raster land cover map,

$$\text{LSI} = \frac{0.25E}{\sqrt{A}} \ .$$

As the LSI increases above 1, the landscape shape becomes more irregular or the length of edge within the landscape increases or both. The *area-weighted mean shape index* is the sum, across all patches, of each patch perimeter divided by the square root of patch area, adjusted by a constant for a square standard, multiplied by the patch area and divided by the total landscape area. To formulate, for $i = 1, \cdots, k$ patch types (land cover categories) and $j = 1, \cdots, n_i$ patches within type i, let p_{ij} and a_{ij} equal the perimeter and area, respectively, for the jth patch of the ith type. Then,

$$\text{AWMSI} = \sum_{i=1}^{k} \sum_{j=1}^{n_i} \left(\frac{0.25 p_{ij}}{\sqrt{a_{ij}}} \right) \left(\frac{a_{ij}}{A} \right).$$

The AWMSI provides an average shape index of patches, weighted by patch area so that large patches are weighted higher than smaller ones. As with the LSI, the patch shapes become more irregular as AWMSI increases above 1.

The perimeter/area scaling exponent, called the double-log fractal dimension (DLFD) by FRAGSTATS, equals 2 divided by the slope of the regression line obtained by regressing the logarithm of patch area against the logarithm of patch perimeter. With limits of 1 and 2, the DLFD reflects more Euclidean shaped patches as DLFD approaches 1 and reflects patches with more convoluted, plane filling perimeters as DLFD approaches 2. The *area-weighted mean patch fractal dimension* equals the sum, across all patches, of 2 times the logarithm of patch perimeter divided by the logarithm of patch area, multiplied by the patch area and divided by the total landscape area. This is expressed as

$$\text{AWMPFD} = \sum_{i=1}^{k} \sum_{j=1}^{n_i} \left(\frac{2\log(0.25 p_{ij})}{\log a_{ij}} \right) \left(\frac{a_{ij}}{A} \right).$$

Shannon evenness was chosen to characterize the marginal land cover distribution, and is simply the Shannon entropy of the land cover proportions divided by the maximum attainable entropy. Therefore, for $i = 1, \cdots, k$ land cover types and P_i equals the proportion of data pixels in the landscape that are categorized as type i,

$$\text{SHEI} = \frac{-\sum_{i=1}^{k} P_i \log P_i}{\log(k)},$$

and as SHEI approaches 0 from above, the landscape is increasingly dominated by particular land cover types, whereas as SHEI approaches 1 from below, the distribution of land cover types becomes increasingly more even.

The *interspersion and juxtaposition* and *Shannon contagion* indices respond to both the composition and configuration of landscape pattern.

For contagion, consider a cell adjacency matrix where each element A_{ij} is the number of pixels in the raster map of category i that are adjacent to category j; then let $v_{ij} = \frac{A_{ij}}{\sum_{i=1}^{k} \sum_{j=i}^{k} A_{ij}}$. Shannon evenness of the adjacency matrix can then be obtained, and it's compliment is taken as a measure of contagion as

$$\text{SHCO} = 1 + \frac{\sum_{i=1}^{k} \sum_{j=i}^{k} v_{ij} \log(v_{ij})}{2 \log(k)} .$$

The interspersion and juxtaposition index is similar to Shannon contagion, except that it quantifies the unevenness of the *patch* adjacency matrix, as opposed to the cell adjacency matrix. For k land cover types, let e_{ij} equal the total length of edge in the landscape between land cover types i and j, and let E equal the total length of edge in the landscape. Interspersion and juxtaposition is then measured by

$$\text{IJI} = -\frac{\sum_{i=1}^{k} \sum_{j=i+1}^{k} \frac{e_{ij}}{E} \log(\frac{e_{ij}}{E})}{\log(\frac{1}{2} k(k-1))} .$$

2.   The Conditional Entropy Profile

Johnson, Tempelman and Patil (1995) introduce conditional entropy as a basis for quantifying spatial fragmentation patterns as one moves from coarser- to finer-resolution categorical raster maps of a landscape. We now present the concept of conditional entropy, followed by a computational approach in the case of multiple raster map resolutions that are obtained by scaling up from a floor resolution by a random filter.

Consider a raster map whose pixels are each assigned one of k distinct land cover categories that are in turn represented by distinct colors. Now consider a map of the same extent that consists of four times as many pixels such that they are hierarchically nested within the larger pixels of the first map. This sequence of increasing resolution may continue so that for an arbitrary n^{th} scale (resolution), each pixel can be sub-divided into four "child" pixels at the $n+1$ scale, and is itself a child pixel of a "parent" pixel from the $n-1$ scale. This process is diagrammed in Figure 2.1.

Letting the most coarse scale map be level 0, and the finest scale (floor resolution) be level L, we thus have a sequence $M_0, \cdots, M_L$ of maps with increasingly finer resolution. For the nth resolution, let $\hat{P}_i^{[n]}$ equal the proportion of pixels from the nth scaled map that are labeled as category i for $i = 1, \cdots, k$, where the "hat" symbolizes a quantity that is calculated from data. For each set of four child pixels at scale $n+1$ that are nestedwithin a single parent pixel, let s be a vector containing

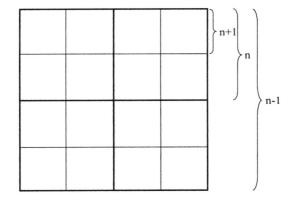

Figure 2.1.   Hierarchical nesting of pixels.

a unique ordering of four out of k categories for $\mathbf{s} = 1, \cdots, k^4$, where the ordering is according to the pixel sequence: NW corner, NE corner, SW corner, SE corner.

Let $\hat{P}_{\mathbf{s}}^{[n+1]}$ equal the proportion of child "4-tuples" that yield the vector $\mathbf{s}$ at scale $n+1$. Now define $\hat{P}_{i\mathbf{s}}^{[n,n+1]}$ as the proportion of "4-tuples" at scale $n+1$ that are of vector $\mathbf{s}$, given a parent pixel of category i, and define $\hat{P}_{\mathbf{s}i}^{[n+1,n]}$ as the proportion of parent pixels in category i, given that a child 4-tuple is of vector $\mathbf{s}$.

For a particular scale n, the marginal entropy of the parent pixels from the n scaled map is computed as

$$\hat{H}^{[n]} = -\sum_{i=1}^{k} \hat{P}_i^{[n]} \log \hat{P}_i^{[n]} , \tag{2.1}$$

the marginal entropy of the child 4-tuples from the $n+1$ scaled map is computed as

$$\hat{H}^{[n+1]} = -\sum_{\mathbf{s}=1}^{k^4} \hat{P}_{\mathbf{s}}^{[n+1]} \log \hat{P}_{\mathbf{s}}^{[n+1]} , \tag{2.2}$$

the conditional entropy of the parent scale categories, given the child scale 4-tuples is computed as

$$\hat{H}^{[n+1,n]} = -\sum_{\mathbf{s}=1}^{k^4} \hat{P}_{\mathbf{s}}^{[n+1]} \sum_{i=1}^{k} \hat{P}_{\mathbf{s}i}^{[n+1,n]} \log \hat{P}_{\mathbf{s}i}^{[n+1,n]} , \tag{2.3}$$

and the conditional entropy of the child scale 4-tuples given the parent
scale categories is computed as

$$\hat{H}^{[n,n+1]} = -\sum_{i=1}^{k} \hat{P}_i^{[n]} \sum_{s=1}^{k^4} \hat{P}_{is}^{[n,n+1]} \log \hat{P}_{is}^{[n,n+1]} , \qquad (2.4)$$

where $x\log(x)$ equals 0 when $x = 0$.

The objective is to obtain this last expression $\hat{H}^{[n,n+1]}$, although it
is difficult to compute directly. However, since the total entropy of
cross-classified factors can be decomposed into among and within sources
(i.e. Patil and Taillie, 1982; Pileou, 1975; Colwell, 1974), the entropy
components just discussed are related as

$$\hat{H}^{[n]} + \hat{H}^{[n,n+1]} = \hat{H}^{[n+1]} + \hat{H}^{[n+1,n]} . \qquad (2.5)$$

Therefore, if the other three entropy components are obtained, as will
be discussed shortly, then $\hat{H}^{[n,n+1]}$ is readily solved for.

This conditional entropy is bound below by zero and above by $\log(k^4)$.
Zero is achieved when, for all $\hat{P}_i^{[n]} > 0$, $\hat{P}_{is}^{[n,n+1]} = 1$ for some particular
s (that may depend on i) and $\hat{P}_{is'}^{[n,n+1]} = 0$ for all $s \neq s'$. The $\log(k^4)$
is achieved when, for all $\hat{P}_i^{[n]} > 0$, the $\hat{P}_{is}^{[n,n+1]}$ are equal for all s and
therefore equal $1/k^4$. In other words, the lower bound is achieved when
each parent pixel can only be subdivided into child pixels of one category,
and the upper bound is achieved when there is an even distribution of
categories among child 4-tuples that are nested within parent pixels of
a common parent category.

To use the terminology of Colwell (1974), the upper bound implies
a state of maximum conditional entropy and zero predictability of the
$n + 1$ scale map, given the n scale map, while the lower bound implies
zero conditional entropy and complete predictability of the $n + 1$ scale
map, given the n scale map. For actual landscapes, it is unlikely to
ever see a maximum conditional entropy of $\log(k^4)$, but rather a lower
maximum that is a function of the marginal land cover distribution.

When conditional entropy is plotted as a function of increasing pixel
size (decreasing resolution), we trace a *conditional entropy profile*, examples of which appear in following chapters. This simple 2-dimensional
plot reflects characteristics of both the marginal land cover distribution
and multi-scale spatial pattern.

2.1   Computing Conditional Entropy of Expected Frequencies based on Single -Resolution Maps

Multi-resolution data for the framework that was just described is
seldom available in practice. More typically, only a single resolution

map is available, referred to as the "floor" resolution. We may obtain a sequence of increasingly coarser maps from this floor resolution map, by successive application of a resampling filter, whereby parent pixels are assigned categories according to some function of its corresponding children pixels.

Several approaches to resampling are available. For example, Costanza and Maxwell (1994) chose the northwest corner child pixel to assign its category to the respective parent pixel. A modal filter (i.e. Benson and Mackenzie, 1995) may also be used, which has intuitive appeal; however, the child pixels can easily not always yield a single mode; therefore, "ties" need to be broken at random anyhow. A modal filter is more appealing for larger resampling windows, such as when aggregating 3x3 or more child pixels into each parent.

A random filter, which chooses one of the child pixels with equal probability, is a close approximation to the modal filter. Most importntly, the random filter allows for computation of expected frequencies $P_i^{[n]} = E[\hat{P}_i^{[n]}]$, $P_s^{[n+1]} = E[\hat{P}_s^{[n+1]}]$ and $P_{s,i}^{[n+1,n]} = E[\hat{P}_{s,i}^{[n+1,n]}]$, as discussed shortly. We can therefore obtain the entropy of these expected frequencies, which is much more meangingful than calculating conditional entropy for one particular simulation of the random filter. Even if we simulated the random filter many times, calculating the conditional entropy for each replication and taking the average, such a sample entropy is well known to be biased (i.e. Basharin, 1959). For these reasons, we chose to work with a random filter so we could obtain the expected frequencies and subsequent entropies, as discussed next. Further computational details and supporting theory are found in Johnson, Myers, Patil and Taillie (1998) and Johnson, Myers, Patil and Taillie (2001a).

By virtue of the random filter, the category (or color) of each pixel at each resolution is a random variable, so we must consider corresponding "color" distributions. For example, a pixel u in map M_n has a color distribution such that $f_u^{[n]}(i)$, for $i = 1, \cdots, k$, is the probability that pixel u is assigned color i. Letting (v_1, v_2, v_3, v_4) be the 4-tuple of child pixels of u in map M_{n+1}, the random filter implies that

$$f_u^{[n]}(i) = \frac{1}{4}(f_{v_1}^{[n+1]}(i) + f_{v_2}^{[n+1]}(i) + f_{v_3}^{[n+1]}(i) + f_{v_4}^{[n+1]}(i)). \qquad (2.6)$$

In other words, when using the random filter, the probability of assigning the ith color to pixel u in map M_n equals the average probability that the ith color is assigned to each child pixel in map M_{n+1} that is nested within pixel u. Therefore, instead of actually applying the random filter to map M_{n+1} to obtain one particular realization of map M_n, an overall distribution of possible colors can be obtained by the *linear filter* on spatially referenced color distributions that is presented by Equation 2.6.

The "floor" resolution map (M_L), which was obtained directly from the data, has a degenerate color distribution for each pixel because the colors are known. In other words, at the floor resolution, $f_v^{[L]}(i) = 1$ if pixel v has color i and $f_v^{[L]}(i) = 0$ otherwise. Starting with the floor resolution, the color distributions at successively coarser resolutions are obtained by recursively applying Equation 2.6.

If we let $\mathbf{s} = (j_1, j_2, j_3, j_4)$ be a 4-tuple of colors, then the probability that the colors $\mathbf{s}$ are assigned to the 4-tuple of pixels $\mathbf{v} = (v_1, v_2, v_3, v_4)$ is

$$f_{\mathbf{v}}^{[n+1]} \equiv f_{v_1}^{[n+1]}(j_1) f_{v_2}^{[n+1]}(j_2) f_{v_3}^{[n+1]}(j_3) f_{v_4}^{[n+1]}(j_4). \qquad (2.7)$$

While equation 2.7 is not true for random maps in general, it is obviously true for the deterministic floor resolution map and holds true for the successively coarser resolution maps because they are obtained by applying the random filter independently in each 2×2 window.

Now we can obtain the expected marginal distribution of 4-tuple colors for the $n + 1$ resolution, $P_{\mathbf{s}}^{[n+1]}$, for $\mathbf{s} = 1, \cdots, k^4$, by averaging the expression in Equation 2.7 over N_n windows $\mathbf{v}$ in map M_{n+1}, such as

$$P_{\mathbf{s}}^{[n+1]} = \frac{1}{N_n} \sum_{\mathbf{v}} f_{v_1}^{[n+1]}(j_1) f_{v_2}^{[n+1]}(j_2) f_{v_3}^{[n+1]}(j_3) f_{v_4}^{[n+1]}(j_4). \qquad (2.8)$$

However, if masking of floor resolution pixels is necessary, we must delete 4-tuples $\mathbf{s}$ containing "nodata" values and renormalize the $P_{\mathbf{s}}^{[n+1]}$ table. Note that N_n is the number of pixels in map M_n that also equals the number of 2×2 non-overlapping windows in map M_{n+1}.

Next we consider the marginal probabilities $P_i^{[n]}$. These probabilities do not depend upon the resolution n when the datamap is square of size $2^L \times 2^L$. If pixel-masking is necessary, then $P_i^{[n]}$ can vary slightly with the resolution n. First, observe that

$$P_i^{[n]} = \frac{1}{N_n} \sum_{u} f_u^{[n]}(i), \qquad (2.9)$$

where N_n is the number of pixels in map M_n and the sum is over all pixels u in map M_n. Again, deletion of the "nodata" value i and renormalization of the table is required in case of pixel-masking. Summing both sides of equation 2.6 with respect to u and using the fact that $4N_n = N_{n+1}$ establishes the earlier claim that $P_i^{[n]}$ does not depend upon n when there is no pixel-masking. The argument breaks down when there is pixel-masking because the sum must exclude the "nodata" values.

Table 2.2. The five types of 4-tuples s and corresponding values of Q_s in equation 2.11, using natural logarithms. The second column gives a canonical permutation of the components of s and the third column is obtained from equation 2.10.

Type	$j_1 j_2 j_3 j_4$	$P_{s,i}^{[n+1,n]}$	Q_s
1	AAAA	δ_{Ai}	$-1 \log 1 = 0$
2a	AAAB	$\frac{3}{4}\delta_{Ai} + \frac{1}{4}\delta_{Bi}$	$-\frac{3}{4}\log\frac{3}{4} - \frac{1}{4}\log\frac{1}{4} = 0.562$
2b	AABB	$\frac{1}{2}\delta_{Ai} + \frac{1}{2}\delta_{Bi}$	$-\frac{1}{2}\log\frac{1}{2} - \frac{1}{2}\log\frac{1}{2} = 0.693$
3	AABC	$\frac{1}{2}\delta_{Ai} + \frac{1}{4}\delta_{Bi} + \frac{1}{4}\delta_{Ci}$	$-\frac{1}{2}\log\frac{1}{2} - \frac{2}{4}\log\frac{1}{4} = 1.04$
4	ABCD	$\frac{1}{4}(\delta_{Ai} + \delta_{Bi} + \delta_{Ci} + \delta_{Di})$	$-\frac{4}{4}\log\frac{1}{4} = 1.386$

Meanwhile, the conditional probability distribution of colors in map M_n, given the respective 4-tuples of colors in M_{n+1}, is readily obtained by the following relationship. For color vector $s=(j_1, j_2, j_3, j_4)$,

$$P_{s,i}^{[n+1,n]} = \frac{1}{4}(\delta_{j_1,i} + \delta_{j_2,i} + \delta_{j_3,i} + \delta_{j_4,i}), \qquad (2.10)$$

where $\delta_{j,i}$ is the kronecker delta function such that

$$\delta_{j,i} = \begin{cases} 1 & \text{if } j = i \\ 0 & \text{otherwise.} \end{cases}$$

The formidable-looking expression,

$$Q_s = -\sum_{i=1}^{K} P_{s,i}^{[n+1,n]} \log(P_{s,i}^{[n+1,n]}), \qquad (2.11)$$

that appears in equation 2.3 is actually quite simple and does not depend on the resolution n or on the floor resolution map. We are going to show that Q_s takes on only five distinct values as s ranges over the K^4 different 4-tuples.

First observe, from equation (2.10), that Q_s is unchanged when the components of $s = (j_1, j_2, j_3, j_4)$ are permuted. The 4-tuples s can be classified into four different types depending on the number of distinct colors among j_1, j_2, j_3, j_4. The 4-tuples of type 2 (i.e., those with exactly two distinct colors) can be further classified into two subtypes depending on whether the two colors are distributed among the four pixels in a 50-50 split or a $\frac{1}{4}$-$\frac{3}{4}$ split. Table 2.2 lists the five different types of 4-tuples s and the corresponding values of Q_s for a natural logarithm. The second column of the table gives a canonical permutation of the components of s for each type.

Equation 2.3 now takes the simple form of a linear combination of the probabilities (with respect to the random filter) of occurrence of the different types of 4-tuples in M_{n+1}:

$$
\begin{aligned}
H^{[n,n+1]} &= \sum_{\mathbf{s}} P_{\mathbf{s}}^{[n+1]} Q_{\mathbf{s}} \\
&= 0.562 \sum_{\mathbf{s}:type2a} P_{\mathbf{s}}^{[n+1]} + 0.693 \sum_{\mathbf{s}:type2b} P_{\mathbf{s}}^{[n+1]} \\
&\quad + 1.04 \sum_{\mathbf{s}:type3} P_{\mathbf{s}}^{[n+1]} + 1.386 \sum_{\mathbf{s}:type4} P_{\mathbf{s}}^{[n+1]}. \quad (2.12)
\end{aligned}
$$

One now has all of the "ingredients" for obtaining the marginal and conditional entropies of the probability distributions $P_i^{[n]}, P_{\mathbf{s}}^{[n+1]}$ and $P_{\mathbf{s},i}^{[n+1,n]}$, which are $H^{[n]}, H^{[n+1]}$ and $H^{[n+1,n]}$, respectively. Therefore, Equation 2.5 can now be used to obtain the conditional entropy of the probability of going to a child 4-tuple of colors $\mathbf{s}$ at scale $n+1$, given the parent color i at scale n, such that

$$
\begin{aligned}
H^{[n,n+1]} &= -\sum_{i=1}^{k} P_i^{[n]} \sum_{\mathbf{s}=1}^{k^4} P_{i\mathbf{s}}^{[n,n+1]} \log P_{i\mathbf{s}}^{[n,n+1]} \\
&= H^{[n+1]} + H^{[n+1,n]} - H^{[n]} . \quad (2.13)
\end{aligned}
$$

We then obtain an explicit expression for conditional entropy by substituting the terms $P_i^{[n]}$ and $P_{\mathbf{s}}^{[n+1]}$ into equations 2.1 and 2.2, respectively, then combining with equations 2.12 and 2.13, resulting in

$$
\begin{aligned}
H^{[n,n+1]} &= -\sum_{\mathbf{s}} P_{\mathbf{s}}^{[n+1]} \log(P_{\mathbf{s}}^{[n+1]}) + \sum_{i=1}^{K} P_i^{[n]} \log(P_i^{[n]}) \\
&\quad + 0.562 \sum_{\mathbf{s}:type2a} P_{\mathbf{s}}^{[n+1]} + \sum_{\mathbf{s}:type2b} P_{\mathbf{s}}^{[n+1]} \\
&\quad + 1.04 \sum_{\mathbf{s}:type3} P_{\mathbf{s}}^{[n+1]} + 1.386 \sum_{\mathbf{s}:type4} P_{\mathbf{s}}^{[n+1]} . \quad (2.14)
\end{aligned}
$$

Chapter 3

ILLUSTRATIONS

The methods for obtaining empirical conditional entropy profiles that were discussed in the last chapter are illustrated in this chapter. We initially start with hypothetical raster maps that are square and contain just 256 pixels at the floor resolution. Each pixel is categorized into black or white, which may be thought of as "forest" or "non-forest", respectively. These initial examples are purposely very simplistic in order to illustrate the calculations.

1. Example 1: Checkerboard Map

Let's first consider the checkerboard map in Figure 3.1. Using natural logarithms, the marginal entropy of the $k = 2$ categories is

$$H^{[n]} = -\frac{1}{2}\log\frac{1}{2} - \frac{1}{2}\log\frac{1}{2} = 0.693,$$

which is actually the maximum possible entropy since the pixels are evenly distributed between black and white. Recall that by virtue of the random filter, this value is constant for square maps with no missing data for all resolutions. Now we obtain the other necessary entropy components, $H^{[n+1]}$ and $H^{[n+1,n]}$, then solve Equation 2.14 for the conditional entropy $H^{[n,n+1]}$ for each parent-child pair of resolutions.

floor+1 → floor:
Since all 4-tuples at the floor resolution are either $\{BBBB\}$ or $\{WWWW\}$, then each will be aggregated up to a black or white parent pixel, respectively (see Figure 3.1A); therefore, the parent pixel categories are uniquely determined and so the entropy associated with going from the floor to the floor + 1 resolution, $H^{[n+1,n]}$, is zero (no uncertainty). Like-

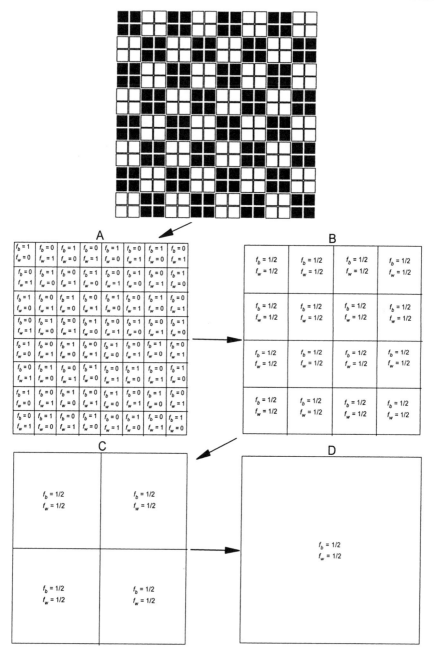

Figure 3.1. Hypothetical checkerboard map at the floor resolution and results from scaling up with the random filter. A = floor+1, B = floor+2, C = floor+3, D = floor+4.

wise, the entropy of the floor resolution 4-tuples is

$$H^{[n+1]} = -\frac{1}{2}\log\frac{1}{2} - \frac{1}{2}\log\frac{1}{2} = 0.693.$$

Therefore,

$$H^{[n,n+1]} = 0.693 + 0 - 0.693 = 0.$$

floor+2 → floor+1:

For the next increase in resolution, the expected frequency of the parent pixels is evenly distributed between black and white (see Figure 3.1B), thus

$$H^{[n+1,n]} = -\frac{1}{2}\log\frac{1}{2} - \frac{1}{2}\log\frac{1}{2} = 0.693.$$

Meanwhile, the child 4-tuples are uniquely determined as $\{BWWB\}$, so $H^{[n+1]} = -1\log(1) = 0$ and therefore

$$H^{[n,n+1]} = 0 + 0.693 - 0.693 = 0.$$

floor+3 → floor+2:

For the next increase in resolution, all $2^4 = 16$ possible child 4-tuples are equally likely and therefore each has a one in sixteen chance of occuring. With respect to Table 2.2, there are 2 possibilities for a *type 1* 4-tuple, 8 possibilities for a *type 2a* 4-tuple and 6 possibilities for a type *type 2b* 4-tuple. Therefore,

$$
\begin{aligned}
H^{[n+1,n]} &= 8 \times \frac{1}{16}\left(-\frac{3}{4}\log\frac{3}{4} - \frac{1}{4}\log\frac{1}{4}\right) \\
&+ 6 \times \frac{1}{16}\left(-\frac{1}{2}\log\frac{1}{2} - \frac{1}{2}\log\frac{1}{2}\right) \\
&= 0.541
\end{aligned}
$$

and

$$H^{[n+1]} = -16\left(\frac{1}{16}\log\frac{1}{16}\right) = 2.773.$$

Therefore,

$$H^{[n,n+1]} = 2.773 + 0.541 - 0.693 = 2.621.$$

floor+4 → floor+3:

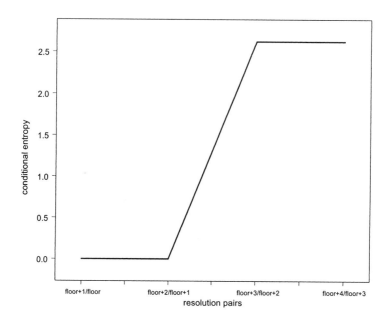

Figure 3.2. Conditional entropy profile corresponding to Example 1, a simple checkerboard pattern.

We see by the expected frequencies in Figure 3.1 (C and D) that the same calculations will apply as in the last step; therefore

$$H^{[n+1,n]} = 0.541, \quad \text{and} \quad H^{[n+1]} = 2.773,$$

so

$$H^{[n,n+1]} = 2.773 + 0.541 - 0.693 = 2.621.$$

A conditional entropy profile is then created by plotting the values of $H^{[n,n+1]}$ against their respective resolutions, as seen in Figure 3.2

2.   Example 2: Irregular Black and White Map

Now let's move to a somewhat more realistic landscape, although still hypothetical and very simplified. Consider the binary map in Figure 3.3. Here we see a more seemingly random placement of pixel categories, with a dominance of the color black. This may be thought of as a forest (black) dominated landscape.

The marginal entropy of the $k = 2$ categories is

$$H^{[n]} = -\frac{184}{256} \log \frac{184}{256} - \frac{72}{256} \log \frac{72}{256} = 0.594,$$

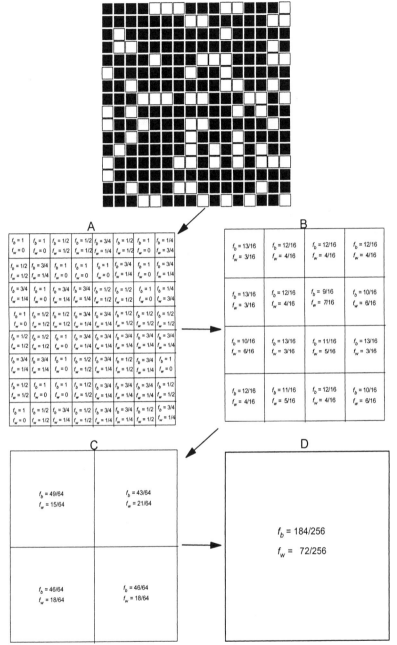

Figure 3.3. Hypothetical binary landscape map at the floor resolution and results from scaling up with the random filter. A = floor+1, B = floor+2, C = floor+3, D = floor+4.

which applies to all resolutions.

floor+1 → floor:

For scaling up from the floor to the next highest resolution, we have

$$
\begin{aligned}
H^{[n+1,n]} &= \left(-\frac{3}{4}\log\frac{3}{4} - \frac{1}{4}\log\frac{1}{4}\right)\left(\frac{26}{64}\right) \\
&+ \left(-\frac{1}{2}\log\frac{1}{2} - \frac{1}{2}\log\frac{1}{2}\right)\left(\frac{22}{64}\right) \\
&= 0.467,
\end{aligned}
$$

Now we need to obtain the floor resolution distribution of 4-tuples, which is seen in Table 3.1. The entropy of this distribution is thus

$$
H^{[n+1]} = -\sum_{s=1}^{16} P_s^{[n+1]} \log P_s^{[n+1]} = 2.206.
$$

Therefore,

$$
H^{[n,n+1]} = 2.206 + 0.467 - 0.594 = 2.079.
$$

Table 3.1. Distribution of 16 possible 4-tuples at the floor resolution, with respect to Example 2.

	BB BB	BB BW	BB WB	BW BB	WB BB	BB WW	WW BB	BW BW	WB WB	BW WB	WB BW	BW WW	WB WW	WW BW	WW WB	WW WW
count	16	7	10	5	3	2	8	3	3	1	5	1	0	0	0	0
proportion	0.25	0.11	0.16	0.08	0.05	0.03	0.13	0.05	0.05	0.02	0.08	0.02	0	0	0	0

floor+2 → *floor*+1:

For the higher resolutions, a spreadsheet can assist in the computations. Table 3.2, which is essentially an enlargement of part A of Figure 3.3, shows the expected frequencies of black and white pixels at the floor+1 resolution after applying Equation 2.6. Since the upper left 4-tuple at the floor resolution is all black, then the expected frequency of a black parent pixel at the floor+1 resolution equals 1; therefore a value of 1.00 is reported in the upper left of Table 3.2 and a value of 0.00 is placed below it, thus completing the expected frequencies for the parent pixel. This is repeated throughout Table 3.2 in a likewise manner. The horizontal and vertical lines in Table 3.2 demarcate the parent pixels at the floor+2 resolution.

Table 3.3 presents the expected 4-tuple frequencies according to Equation 2.8. Here, each row presents the expected distribution of 4-tuples at the floor+1 resolution that correspond with the respective parent pixel at the floor+2 resolution. The parents are listed according to their (row, column) location in Table 3.2. Applying Equation 2.8 is the same as averaging the columns across all rows in Table 3.3, resulting in the last row in Table 3.3.

Using the 4-tuple distribution in Table 3.3,

$$
\begin{aligned}
H^{[n+1,n]} &= \left(-\frac{3}{4}\log\frac{3}{4} - \frac{1}{4}\log\frac{1}{4}\right) \times 0.50 \\
&+ \left(-\frac{1}{2}\log\frac{1}{2} - \frac{1}{2}\log\frac{1}{2}\right) \times 0.25 \\
&= 0.454
\end{aligned}
$$

and

$$
H^{[n+1]} = -\sum_{s=1}^{16} P_s^{[n+1]}\log P_s^{[n+1]} = 2.342,
$$

therefore

$$
H^{[n,n+1]} = 2.342 + 0.454 - 0.594 = 2.202.
$$

floor+3 → *floor*+2:

In line with the discussion above, the top of Table 3.4 is equivalent to part B of Figure 3.3 and the bottom of Table 3.4 provides the distribution of 4-tuples. The corresponding entropy components are thus

$$
H^{[n+1,n]} = \left(-\frac{3}{4}\log\frac{3}{4} - \frac{1}{4}\log\frac{1}{4}\right) \times 0.48
$$

$$+ \left(-\frac{1}{2}\log\frac{1}{2} - \frac{1}{2}\log\frac{1}{2}\right) \times 0.24$$

$$= 0.436$$

and

$$H^{[n+1]} = 2.346.$$

Therefore,

$$H^{[n,n+1]} = 2.346 + 0.436 - 0.594 = 2.188.$$

floor+4 → *floor+3*:

Now, the top of Table 3.5 is equivalent to part C of Figure 3.3 and the bottom of Table 3.5 provides the distribution of 4-tuples. The corresponding entropy components are thus

$$H^{[n+1,n]} = \left(-\frac{3}{4}\log\frac{3}{4} - \frac{1}{4}\log\frac{1}{4}\right) \times 0.48$$

$$+ \left(-\frac{1}{2}\log\frac{1}{2} - \frac{1}{2}\log\frac{1}{2}\right) \times 0.25$$

$$= 0.443$$

and

$$H^{[n+1]} = 2.366.$$

Therefore,

$$H^{[n,n+1]} = 2.366 - 0.594 + 0.443 = 2.215.$$

Table 3.2. Expected color frequencies corresponding to the floor+1 resolution. The horizontal and vertical lines demarcate parent pixels at the floor+2 resolution.

b	1.00	1.00	0.50	0.50	0.75	0.50	1.00	0.25
w	0.00	0.00	0.50	0.50	0.25	0.50	0.00	0.75
b	0.50	0.75	1.00	1.00	1.00	0.75	1.00	0.75
w	0.50	0.25	0.00	0.00	0.00	0.25	0.00	0.25
b	0.75	1.00	0.75	0.75	0.50	0.50	1.00	0.50
w	0.25	0.00	0.25	0.25	0.50	0.50	0.00	0.50
b	1.00	0.50	0.50	0.75	0.75	0.50	0.50	0.50
w	0.00	0.50	0.50	0.25	0.25	0.50	0.50	0.50
b	0.50	0.50	1.00	0.75	0.75	0.75	0.75	0.75
w	0.50	0.50	0.00	0.25	0.25	0.25	0.25	0.25
b	0.75	0.75	1.00	0.50	0.75	0.50	0.75	1.00
w	0.25	0.25	0.00	0.50	0.25	0.50	0.25	0.00
b	0.50	1.00	1.00	0.50	0.75	0.75	0.75	0.50
w	0.50	0.00	0.00	0.50	0.25	0.25	0.25	0.50
b	1.00	0.50	0.75	0.50	0.75	0.75	0.50	0.75
w	0.00	0.50	0.25	0.50	0.25	0.25	0.50	0.25

Table 3.3. Distribution of 16 possible 4-tuples at the floor+1 resolution, with respect to Example 2.

parent (row,column)	BB BB	BB BW	BB WB	BW BB	WB BB	BB WW	WW BB	BW BW	WB WB	BW WB	WB BW	BW WW	WB WW	WW BW	WW WB	WW WW
1,1	0.38	0.13	0.38	0.00	0.00	0.13	0.00	0.00	0.00	0.00	0.00	0.00	0.00	0.00	0.00	0.00
1,2	0.25	0.00	0.00	0.25	0.25	0.00	0.25	0.00	0.00	0.00	0.00	0.00	0.00	0.00	0.00	0.00
1,3	0.28	0.09	0.00	0.28	0.09	0.00	0.09	0.09	0.00	0.00	0.03	0.00	0.00	0.03	0.00	0.00
1,4	0.19	0.06	0.00	0.56	0.00	0.00	0.00	0.19	0.00	0.00	0.00	0.00	0.00	0.00	0.00	0.00
2,1	0.38	0.38	0.00	0.00	0.13	0.00	0.00	0.00	0.00	0.00	0.13	0.00	0.00	0.00	0.00	0.00
2,2	0.21	0.07	0.21	0.07	0.07	0.07	0.02	0.02	0.07	0.07	0.02	0.02	0.02	0.01	0.02	0.01
2,3	0.09	0.09	0.03	0.09	0.09	0.03	0.09	0.09	0.03	0.03	0.09	0.03	0.03	0.09	0.03	0.03
2,4	0.13	0.13	0.13	0.13	0.00	0.13	0.00	0.13	0.00	0.13	0.00	0.13	0.00	0.00	0.00	0.00
3,1	0.14	0.05	0.05	0.14	0.14	0.02	0.14	0.05	0.05	0.05	0.05	0.02	0.02	0.05	0.05	0.02
3,2	0.38	0.38	0.00	0.13	0.00	0.00	0.00	0.13	0.00	0.00	0.00	0.00	0.00	0.00	0.00	0.00
3,3	0.21	0.21	0.07	0.07	0.07	0.07	0.02	0.07	0.02	0.02	0.07	0.02	0.02	0.02	0.01	0.01
3,4	0.42	0.00	0.14	0.14	0.14	0.00	0.05	0.00	0.05	0.05	0.00	0.00	0.00	0.00	0.02	0.00
4,1	0.25	0.25	0.00	0.00	0.25	0.00	0.00	0.00	0.00	0.00	0.25	0.00	0.00	0.00	0.00	0.00
4,2	0.19	0.19	0.06	0.19	0.00	0.06	0.00	0.19	0.00	0.06	0.00	0.06	0.00	0.00	0.00	0.00
4,3	0.32	0.11	0.11	0.11	0.11	0.04	0.04	0.04	0.04	0.04	0.04	0.01	0.01	0.01	0.01	0.00
4,4	0.14	0.05	0.14	0.14	0.05	0.05	0.05	0.05	0.05	0.14	0.02	0.05	0.02	0.02	0.05	0.02
ave.	0.25	0.14	0.08	0.14	0.09	0.04	0.05	0.06	0.02	0.04	0.04	0.02	0.01	0.01	0.01	0.01

Table 3.4. top: Expected color frequencies corresponding to the floor+2 resolution.
bottom: Distribution of 16 possible 4-tuples at the floor+2 resolution, with respect to Example 2.

b	0.8125	0.7500	0.7500	
w	0.1875	0.2500	0.2500	
b	0.8125	0.7500	0.5625	0.6250
w	0.1875	0.2500	0.4375	0.3750
b	0.6250	0.8125	0.6875	0.8125
w	0.3750	0.1875	0.3125	0.1875
b	0.7500	0.6875	0.7500	0.6250
w	0.2500	0.3125	0.2500	0.3750

parent	BB BB	BB BW	BB WB	BW BB	WB BB	BB WW	WW BB	BW BW	WB WB	BW WB	WB BW	BW WW	WB WW	WW BW	WW WB	WW WW
1,1	0.37	0.12	0.09	0.12	0.09	0.03	0.03	0.04	0.02	0.03	0.03	0.01	0.01	0.01	0.01	0.00
1,2	0.20	0.12	0.15	0.07	0.07	0.09	0.02	0.04	0.05	0.05	0.04	0.03	0.03	0.01	0.02	0.01
2,1	0.26	0.12	0.09	0.06	0.16	0.04	0.04	0.03	0.05	0.02	0.07	0.01	0.02	0.02	0.01	0.01
2,2	0.26	0.16	0.09	0.06	0.12	0.05	0.03	0.04	0.04	0.02	0.07	0.01	0.02	0.02	0.01	0.01
ave.	0.27	0.13	0.10	0.08	0.11	0.05	0.03	0.04	0.04	0.03	0.05	0.02	0.02	0.01	0.01	0.01

Table 3.5. top: Expected color frequencies corresponding to the floor+3 resolution. bottom: Distribution of 16 possible 4-tuples at the floor+3 resolution, with respect to Example 2.

b	0.77	0.67
w	0.23	0.33
b	0.72	0.72
w	0.28	0.28

parent	BB BB	BB BW	BB WB	BW BB	WB BB	BB WW	WW BB	BW BW	WB WB	BW WB	WB BW	BW WW	WB WW	WW BW	WW WB	WW WW
1,1	0.27	0.10	0.10	0.13	0.08	0.04	0.04	0.05	0.03	0.05	0.03	0.02	0.01	0.02	0.02	0.01

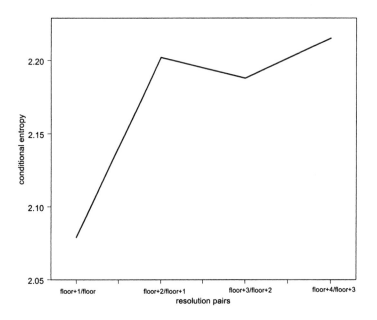

Figure 3.4. Conditional entropy profile corresponding to Example 2, a somewhat random pattern dominated by black pixels.

A conditional entropy profile is then created by plotting the values of $H^{[n,n+1]}$ against their respective resolutions, as seen in Figure 3.4

3.   Example 3: Illustration with Actual Landscapes

The previous examples were purposely simplistic in order to demonstrate how the equations for obtaining conditional entropy are applied. We now step up to a real situation whereby three different watershed-delineated landscapes are chosen to represent a wide range of forest fragmentation, as shown in Figure 3.5. These data are from the statewide land cover map of Pennsylvania that is described in Section 4.1 of Chapter 1.

Tionesta Creek, located in Pennsylvania's northern tier, is a mostly forested watershed that represents a continuum of forest interior wildlife habitat. Jacobs Creek, located south of the city of Pittsburgh, represents a more transitional landscape where the forest is largely fragmented. Meanwhile, the Jordan Creek watershed represents a further fragmented landscape, which barely maintains a connected forest matrix that is encroached by agriculture and urban/suburban land use associated with the Allentown-Bethlehem-Easton metropolitan area.

A computer algorithm has been created for calculating conditional entropy profiles for real landscapes that are delineated by irregular boundaries such as a watershed extent, and are represented by many pixels that are distributed among multiple land cover types (Johnson, *et al.* 2001a). This algorithm was applied to the three watershed-delineated landscapes in Figure 3.5, and the results are seen in Figure 3.6. The profiles in Figure 3.6 "rise" as the forest becomes increasingly fragmented, thus showing how the conditional entropy profiles respond to changing landscape patterns.

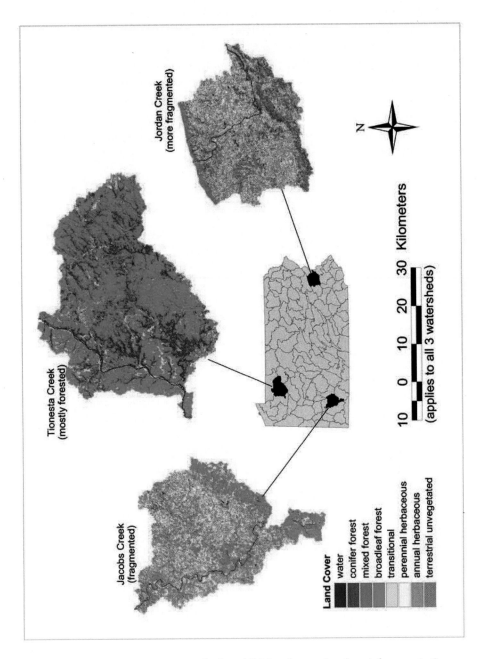

Figure 3.5. Demonstration watersheds exhibiting increasing forest fragmentation.

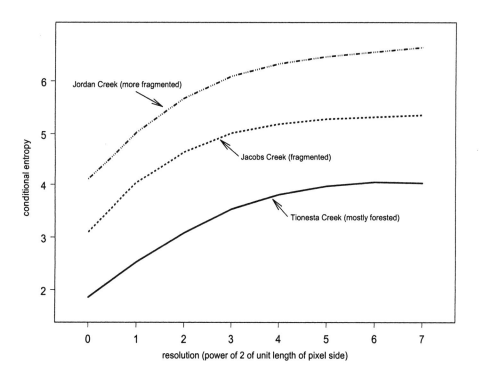

Figure 3.6. Conditional entropy profiles obtained for the three watersheds in Figure 3.5.

Chapter 4

CLASSIFYING PENNSYLVANIA WATERSHEDS ON THE BASIS OF LANDSCAPE CHARACTERISTICS

1.   Introduction

The variables discussed in Chapter 2 were measured for the Pennsylvania watershed-delineated landscapes described in Chapter 1. Along with the single-resolution variables, conditional entropy profiles were also obtained. For these watersheds, a common shape was observed for the profiles, which could be modeled by a negative exponential function

$$y = C - A \times \exp(-B \times x), \tag{4.1}$$

where y equals the conditional entropy and x equals the resolution, represented by the power of 2 of the unit length of a pixel side. In other words, a value of $x = 0$ represents the floor resolution, $x = 1$ represents a doubling of the floor resolution pixels, $x = 2$ represents a quadrupling, etc. Here, the parameters B and A measure the rate and overall amount, respectively, of information loss and the parameter C is the asymptotic maximum attainable conditional entropy.

This is illustrated in Figure 4.1 for the Conestoga Creek Watershed in southeast Pennsylvania.

Note that for a fixed value of B, the transformation $z = exp(-B \times x)$ allows linearization of Equation 4.1 such that $y = C - A \times z$; therefore, for fixed B, ordinary least squares can be applied to estimate values for C and A. Now, for estimated values of C and A, if the residual sum of squares are plotted as a function of changing values of B, one can observe a global minimum, thus providing a least squares estimate for B (Rodríguez, 2001).

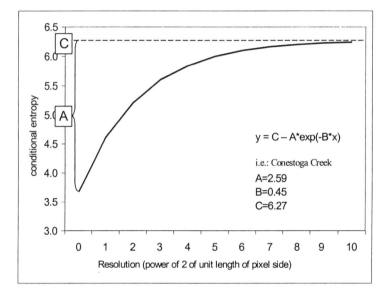

Figure 4.1. Example conditional entropy profile and associated parameters.

With the conditional entropy profiles and other landscape measurements in hand, a primary

objective is to determine if these watershed-delineated landscapes can be classified into common types, and if so, to determine a minimum set of necessary measurements for effectively discriminating among the landscape types. This objective is further refined as follows: first, it is desired to determine whether landscape *pattern* is informative in addition to simple land cover proportions for watersheds of this scale; second, if pattern is important, it is desired to see if the addition of profile variables can improve the ability of single-resolution variables to sensibly categorize the watersheds; third,

it is desired to see if conditional entropy profiles alone, or in conjunction with the readily-available land cover proportions, can do a reasonable job of discriminating among different landscape types. This chapter, based on Johnson, Myers, Patil and Taillie (2001b), attempts to answer these questions for the Pennsylvania watersheds.

All analyses in this and subsequent chapters incorporate the effect of different physiographic regions because both topography and geology influence human activity, which in turn affects landscape pattern. In this chapter, landscape pattern is studied separately for each of the three major physiographic provinces that are discussed in Chapter 1, namely the *Appalachian Plateaus*, the *Ridge and Valley* and the *Piedmont Plateau.*

Smaller provinces, such as the Coastal and Lake Erie Plains and the Blue Ridge Province are ignored because they are too small to fully encompass any of the watersheds used to delineate landscapes in this study.

2. Clustering Watersheds into Common Groups

Relationships among the initial set of single-resolution spatial pattern variables from Table 2.1 are seen in Figure 4.2 for all of the watersheds. The three land cover variables and regression estimates of the profile variables A, B and C are plotted in Figure 4.3. An approximately uncorrelated subset of the pattern variables from Figure 4.2 was added to the plot in 4.3 to visualize how the land covers and profile variables are related to the larger set of pattern variables.

Different subsets of the variables

in Figures 4.2 and 4.3, were used to cluster the watersheds in multivariate space by average Euclidean distance after standardizing the variables. Manhattan distance made little difference once the variables were standardized, and the average linkage protocol was chosen over others for reasons of consistency and robustness. Also, model-based clustering was avoided because it assumes one has independent sample units (MathSoft, Inc. 1997) and the watersheds of this study are a *population* of units that also exhibit high spatial autocorrelation.

The different combinations of variables that were used for clustering the watersheds are listed in table 4.1. The objectives of clustering were twofold. First, it was desired to determine what group of variables yield the most distinct clusters that are readily interpretable with respect to landscape disturbance, based on visual inspection of the land cover maps for the watersheds. In other words, "For watersheds within a single cluster, as obtained from the dendrogram, do the maps appear to have similar land cover patterns, and are these patterns different from those of watersheds in other clusters?"

If the "best" clustering results arose from a large set of variables, then the second objective was to determine if a smaller subset could yield clusters that were not much different. At one extreme, it was desired to test whether or not land cover proportions alone could suffice for obtaining clear, sensible clusters with respect to landscape disturbance. Land cover proportions are expected to be very important for discriminating among watersheds; however, we wanted to know if spatial pattern variables could substantially improve the cluster results. Further, when improvement was possible, we wanted to determine if the conditional entropy profile variables could account for most of the improvement.

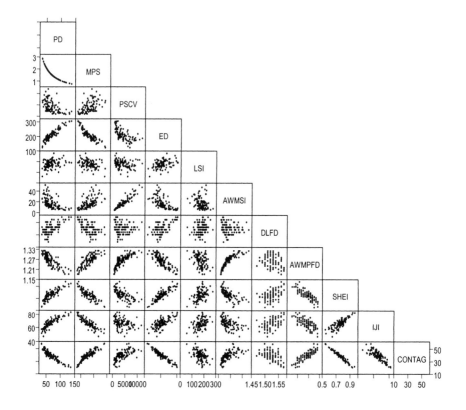

Figure 4.2. Pairwise scatterplots of all watersheds for the landscape pattern variables in Table 2.1.

The best clusters resulted from including all of the variables (group 1 in Table 4.1), regardless of some very high redundancy seen in Figures 4.2 and 4.3.

The resulting dendrograms had very coherent structure that allowed ready separation of watersheds into distinct clusters. Watersheds of a common cluster had very similar land cover patterns that were visually distinct from patterns for other clusters. This resulted in labeling of the clusters by the degree and type of forest fragmentation. The resulting dendrograms appear in Figures 4.4 to 4.6 and the clusters are mapped in Figure 4.7. Qualitative labeling of the clusters ranges from "high" for mostly forested watersheds, to "very low" for watersheds that are primarily agricultural with urban centers and extensive suburban develop-

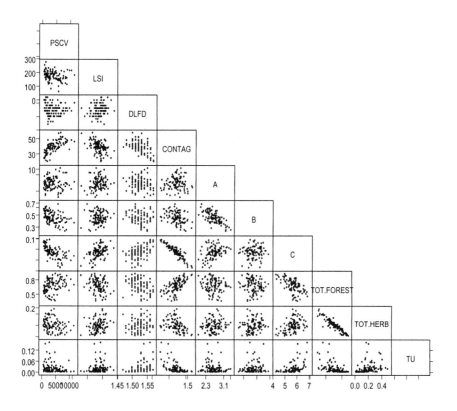

Figure 4.3. Pairwise scatterplots of all watersheds for an approximately orthogonal subset of pattern variables from Figure 4.2, along with the entropy profile variables *A*, *B* and *C*, and the land cover proportions listed in Table 2.1.

ment. These watersheds actually appear worse than those encompassing the cities of Philadelphia and Pittsburgh.

Clustering on the other groups of variables (groups 2 to 5 in Table 4.1) resulted in general classifications that are reported in Tables 4.2 to 4.4 where the results are compared to those obtained using the full set of variables. Resulting dendrograms for groups 2 to 5 in Table 4.1 are not reported; however, the key findings are summarized below.

- For the Piedmont Plateau, land cover proportions alone are sufficient to group the watersheds into clearly distinguishable landscape types. The only difference when clustering was done only with the three land cover proportions, compared to when all variables were

Table 4.1. Different combinations of landscape variables used for clustering Pennsylvania watersheds. All except the last group include the land cover proportions

group	description
1	all variables (11 single-resolution measurements, the profile variables A, B and C and 3 land cover proportions – total forest, total herbaceous and terrestrial unvegetated land)
2	all variables, excluding the profile variables A, B and C
3	just A, B, C and the 3 land covers
4	just the 3 land covers
5	just the spatial pattern variables from group 1 (no land cover)

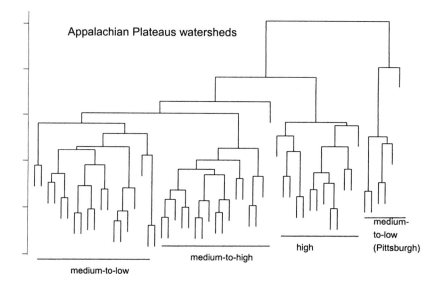

Figure 4.4. Cluster dendrogram of the Appalachian Plateaus watersheds, using the full set of landscape variables. Qualitative labeling of the clusters pertains to amount of forestation.

used, is that White Clay Creek was drawn into the "medium (suburban Philadelphia)" group, which is entirely reasonable. When all the pattern variables were included, but the land cover proportions

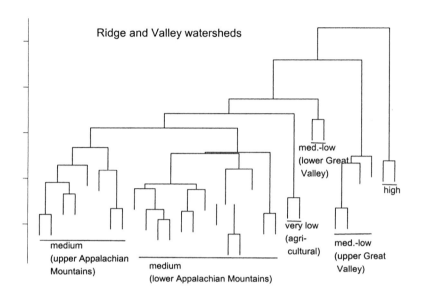

Figure 4.5. Cluster dendrogram of the Ridge and Valley watersheds, using the full set of landscape variables. Qualitative labeling of the clusters pertains to amount of forestation.

were excluded, the "medium" and "very low" watersheds were mixed together.

- For the Ridge and Valley, both land cover proportions and spatial pattern variables were necessary to maintain reasonable clustering. It is noteworthy that when just the land cover proportions were used for clustering, two watersheds (Conodoguinet Creek and Conococheague Creek) were distinctly separated off into their own group prior to splitting the remaining watersheds into clusters. These two watersheds are very similar to the agricultural-dominated watersheds of the "very low" group in the Piedmont Plateau, which were clustered into this distinct group using only land cover.

- For the Ridge and Valley, when only the conditional entropy variables A, B and C were included with the land cover proportions (no single-resolution pattern variables), the resulting clusters were very similar to those obtained with the full set of variables. Some minor

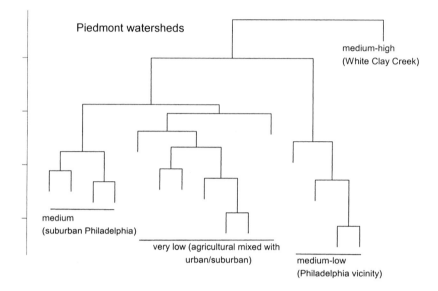

Figure 4.6. Cluster dendrogram of the Piedmont Plateau watersheds, using the full set of landscape variables. Qualitative labeling of the clusters pertains to amount of forestation.

re-shuffling of a few watersheds occured, but nothing that yielded unreasonable results.

When only the single-resolution FRAGSTAT variables were included with the land cover proportions (no ABC), the upper and lower Appalachian Mountains clusters were broken up. Therefore, one can argue that for the Ridge and Valley, just adding the conditional entropy values A, B and C to the land cover proportions may actually perform better than just adding the more conventional single-resolution variables.

- For the Appalachian Plateaus, the spatial pattern variables appear to have a stronger influence than the land cover proportions, precisely the opposite of findings from the Piedmont.

- For the Appalachian Plateaus, when only the conditional entropy variables A, B and C were included with the land cover proportions, one watershed (Turtle Creek) was distinctly separated into its own

Watershed Clusters Based on 11 Single-Resolution Pattern Variables, Conditional Entropy Profile Variables and 3 Land Cover Summaries

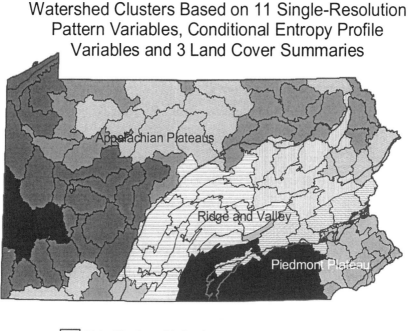

☐ Major Physiographic Provinces

Watershed Clusters

AP high
AP medium-high
AP medium-low
AP medium-low (Pittsburgh)
Piedmont medium (suburban Philadelphia)
Piedmont medium-high (White Clay Creek)
Piedmont medium-low (Philadelphia vicinity)
Piedmont very low (mostly agriculture mixed with urban/suburban)
R&V high
R&V medium (lower Appalachian Mountains)
R&V medium (upper Appalachian Mountains)
R&V medium-low (lower Great Valley)
R&V medium-low (upper Great Valley)
R&V very low (mostly Agriculture)

Figure 4.7. Mapping of the clusters depicted in the dendrograms of Figures 4.4 to 4.6. Cluster labels identify physiographic province membership (AP = Appalachian Plateaus, Piedmont = Piedmont Plateau, R&V = Ridge and Valley) and further geographic information for spatially contiguous clusters. Qualitative statements associated with cluster labels pertain to the relative amount of forestation observed on land cover maps.

Table 4.2. Watershed clustering in the Appalachian Plateaus using all landscape variables, compared to using subsets of variables as indicated. Values in the Table are number of watersheds in each cross-classification.

All Variables	no ABC				
	high[†]	MH	ML	ML(Pitt.)[*]	total
high	9	4	0	0	13
MH	0	17	0	0	17
ML	0	1	18	0	19
ML(Pitt.)	0	0	2	4	6
total	9	22	20	4	55

All Variables	no FRAGSTATS				
	high	MH	ML	Pitt.	total
high	6	7	0	0	13
MH	0	15	2	0	17
ML	0	4	15	0	19
ML(Pitt.)	0	0	5	1	6
total	6	26	22	1	55

All Variables	land cover only				
	high	ML	M(mixed)	ML	total
high	8	5	0	0	13
MH	0	13	3	1	17
ML	0	4	2	13	19
ML(Pitt.)	0	0	3	3	6
total	8	22	8	17	55

All Variables	Pattern only (No Land Cover)				
	high	MH	ML	ML(Pitt.)	total
high	13	0	0	0	13
MH	4	13	0	0	17
ML	1	0	18	0	19
ML(Pitt.)	0	0	0	6	6
total	18	13	18	6	55

[†] Qualitative cluster labels reflect amount of forestation:
MH = medium high, M = medium, ML = medium low
[*] Pitt. = Pittsburgh and near vicinity

group that was very distant (in pattern space) from all other watersheds. This is reasonable because Turtle Creek is unique in that it encompasses the city of Pittsburgh, which is by far the largest urban center in the mostly rural Appalachian Plateaus. However, all of the remaining watersheds were basically split into "medium-to-high" and "medium-to-low" clusters that substantially sacrifices the greater detail obtained from clustering with all the original variables.

Table 4.3. Watershed clustering in the Ridge and Valley using all landscape variables, compared to using subsets of variables as indicated. Values in the Table are number of watersheds in each cross-classification.

All Variables	no ABC					
	high	M†(low AM*)	M(up AM)	ML(up GV)	very low	total
high	1	1	0	0	0	2
M(low AM)	0	13	0	0	0	13
M (up AM)	0	0	8	0	0	8
ML(low GV)	0	1	0	1	0	2
ML (up GV)	0	0	0	4	0	4
very low	0	0	0	0	2	2
total	1	15	8	5	2	31

All Variables	no FRAGSTATS				
	high	M(low AM)	M(up AM)	ML(GV)	total
high	2	0	0	0	2
M (low AM)	0	13	0	0	13
M(up AM)	0	2	6	0	8
ML(low GV)	1	0	0	1	2
ML(up GV)	0	0	0	4	4
very low	0	0	0	2	2
total	3	15	6	7	31

All Variables	3 Land Covers only					
	high	M1	M2	ML	very low	total
high	2	0	0	0	0	2
M(low AM)	0	5	8	0	0	13
M(up AM)	3	0	4	1	0	8
ML(low GV)	0	2	0	0	0	2
ML(up GV)	0	0	0	4	0	4
very low	0	0	0	0	2	2
total	5	7	12	5	2	31

All Variables	no Land Cover						
	high	M1	M2	M3	M4	ML1	ML2	total
high	2	0	0	0	0	0		2
M (low AM)	0	6	6	1	0	0		13
M(up AM)	0	0	0	3	5			8
ML (low GV)	0	0	0	0	0	2		2
ML (up GV)	0	0	0	0	0	0	4	4
very low	0	0	0	2	0	0	0	2
total	2	6	6	6	5	2	4	31

† M = medium and ML = medium low forestation

AM = Appalachian Mountains and GV = Great Valley

Table 4.4. Watershed clustering in the Piedmont using all landscape variables, compared to using subsets of variables as indicated. Values in the Table are number of watersheds in each cross-classification.

All Variables	no ABC				
	MH[†]	M	ML	very low	total
MH	1	0	0	0	1
M	0	4	0	0	4
ML	0	0	4	0	4
very low	1	0	0	6	7
total	2	4	4	6	16

All Variables	no FRAGSTATS			
	M	ML	very low	total
MH	1	0	0	1
M	4	0	0	4
ML	1	3	0	4
very low	0	0	7	7
total	6	3	7	16

All Variables	3 Land Covers only			
	M	ML	very low	total
MH	1	0	0	1
M	4	0	0	4
ML	0	4	0	4
very low	0	0	7	7
total	5	4	7	16

All Variables	No Land Cover				
	MH	M	M and very low	ML	total
MH	1	0	0	0	0
M	0	2	2	0	4
ML	0	0	0	4	4
very low	0	0	7	0	7
total	1	2	7	4	16

[†] symbols for relative amount of forestation are as in Table 4.2

When only the FRAGSTAT variables were included with the land cover proportions (no *ABC*), the resulting clusters were very similar to the original clusters obtained from using all the original variables. Therefore, one can argue that for the Appalachian Plateaus, the single-resolution variables out-perform the conditional entropy profile variables.

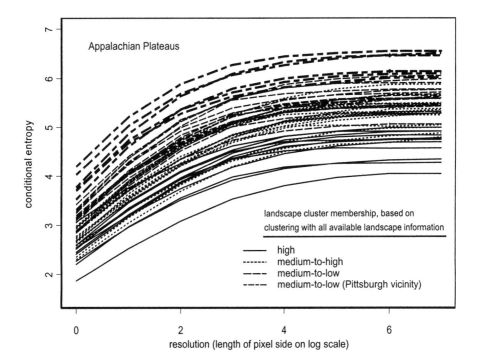

Figure 4.8. Conditional entropy profiles of watersheds in the Appalachian Plateaus physiographic province, coded according to membership in landscape cluster.

- One overall conclusion is that each physiographic province yielded its own unique properties.

3.   Comparison to Conditional Entropy Profiles

Conditional entropy profiles are reported in Figures 4.8 to 4.10 where they are coded with respect to the landscape clusters deciphered in Figures 4.4 to 4.6, respectively, and mapped in Figure 4.7. These profiles were obtained by applying Equation 2.13 to the land cover raster maps for each watershed.

An initial, encouraging observation is that the profiles do separate from each other, thus indicating that they are responsive to changing landscape pattern in a way that can be readily graphed. This achieves a goal that is similar to plotting different landscapes in "pattern space", as

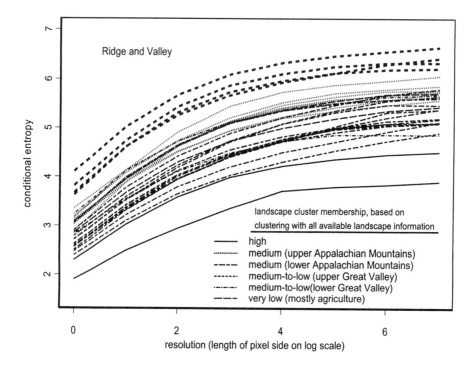

Figure 4.9. Conditional entropy profiles of watersheds in the Ridge and Valley physiographic province, coded according to membership in landscape cluster.

proposed by O'Neill, *et al.* (1996), which allows the ability to measure a landscape's "distance" from some reference. Instead of deciding on a proper set of variables to describe the pattern space of a landscape, the conditional entropy profiles can be used to graphically monitor change in a way that also has the benefit of being inherently multi-scale.

For the Appalachian Plateaus, a clear pattern is seen whereby the profiles rise to the top as the forest becomes increasingly fragmented. For both the Ridge and Valley and Piedmont Plateau, a pattern emerges whereby the profiles rise to the top, then fall to a more contagious state as the overall amount of non-forest land and the size of non-forest patches increases. This is in line with expectation based on modeling results found elsewhere (Johnson, Myers, Patil and Taillie, 1999 and 2001a). Apparently, the most fragmented watersheds of the Appalachian

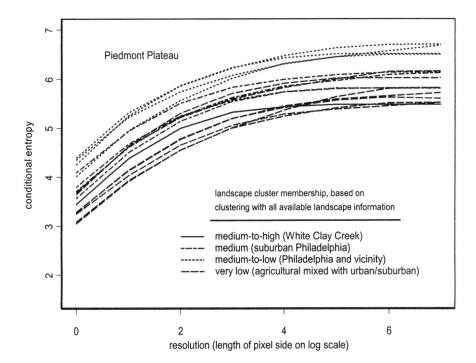

Figure 4.10. Conditional entropy profiles of watersheds in the Piedmont physiographic province, coded according to membership in landscape cluster.

Plateaus are similar to more transitional (medium-to-low) watersheds of either the Piedmont or Ridge and Valley. It is interesting to note that the proportion of total forest cover for the highest overall profile in each of the three provinces is 0.50 for the Ridge and Valley (Jacobs Creek), 0.55 for the Piedmont (Wissahickon Creek) and 0.56 for the Appalachian Plateaus (Chartiers Creek). This indicates that there is some consistency between the proportion of total forest cover and the maximum conditional entropy profile.

These profiles can also be used to compare fragmentation patterns over multiple resolutions between any two landscapes. If a profile from landscape L1 yields a higher conditional entropy than landscape L2 at every resolution, then we may say that landscape L1 is "intrinsically more fragmented" than landscape L2 with respect to the conditional entropy profiles.

Chapter 5

PREDICTABILITY OF SURFACE WATER POLLUTION IN PENNSYLVANIA USING WATERSHED-BASED LANDSCAPE MEASUREMENTS

1. Introduction

This chapter, based on the work by Johnson, Myers and Patil (2001) evaluates the relationship between surface water pollution and landscape characteristics in Pennsylvania watersheds. The objective is to determine if landscape measurements can reliably predict water quality, and if so, to determine the relative contribution of marginal land cover proportions and spatial pattern variables to these predictions.

Two different response variables were obtained from previous studies (Nizeyimana, *et al.*, 1997 and Hamlett, *et al.*, 1992) to assess surface water pollution. The first response variable was total nitrogen loading estimated from in-stream samples. The second was a GIS-modeled pollution potential index. After describing these two response variables, both graphical and principal components analyses are applied to select a set of candidate landscape predictors. Finally, for each of the two response variables, stepwise linear model building is applied to the set of potential predictors in order to choose an "optimal" subset for predicting water quality.

2. Surface Water Pollution Assessment

2.1 Nitrogen Loading

A study by Nizeyimana, *et al.* (1997) was conducted to assess the surface-water nutrient loading, namely nitrogen (N) and phosphorus (P), within select watersheds of Pennsylvania. The primary purpose was to quantify the various sources of non-point source (NPS) nutrient loading. Watersheds, as seen in the top of Figure 5.1, were delineated by choosing 85 Water Quality Network stations throughout Pennsylvania, then aggregating detailed sub-watershed boundaries digitized by the United States Geologic Survey (USGS). Each resulting NPS watershed then drains to one of the 85 network stations. As part of this study, total levels of both nitrogen and phosphorus were obtained for these 85 watersheds by applying "flow-weighted averaging" techniques to monthly in-stream concentrations from the previous 5 years. The result is an estimate of the monthly exported mass in kilograms (kg).

The landscape variables discussed in earlier chapters were calculated for a different set of watersheds (based on the State Water Plan), as shown in the bottom of Figure 5.1. We see that there is not a perfect matchup between the two sets of watersheds. Those "NPS" watersheds that can be aggregated to yield a state water plan watershed are identified in the top of Figure 5.2. These aggregated NPS watersheds are then added to those for which there is an exact or very close match with the state water plan watersheds and the resulting set appears in the bottom of Figure 5.2. The result is a collection of 30 watersheds across the state for which we have both landscape pattern values and nutrient loading values.

For the NPS watersheds that were aggregated, nutrient loading was summed across the sub-watersheds to obtain an aggregate value. All watershed-based estimates of total nitrogen and total phosphorus, in kilograms (kg), were divided by the total area, in hectares (ha), in order to adjust for the varying sizes of the watersheds. Nitrogen was then plotted against phosphorus, as seen in Figure 5.3. Clearly, one only needs to study either nitrogen or phosphorus as an indicator of nutrient loading since they are so highly linearly correlated with each other. Nitrogen was chosen for this purpose because nitrogen loading is always reported as well above zero (minimum for these 30 watersheds = 27.12 kg/ha), whereas phosphorus loading is sometimes reported at less than 1 kg/ha, thus suggesting the potential for more serious analytical detection limit problems for phosphorous.

Since the objective of this study is to evaluate the effect of land use patterns on surface-water nutrient loading, we considered subtracting the portion of total nitrogen loading estimated by Nizeyimana, *et al.* (1997) attributable to atmospheric deposition. However, of the two primary components of atmospheric nitrogen, ammonium (NH_4) was determined to come almost entirely from volatilization from manure and other fertilizers; while the other primary component, nitrogen oxides (NO_x) was determined to have about one third contributed by manure and other fertilizers and about two thirds from industrial/urban sources. Also, natural sources of atmospheric deposition of nitrogen was considered negligible (Nizeyimana, *et al.*, 1997). Since much of the atmospheric

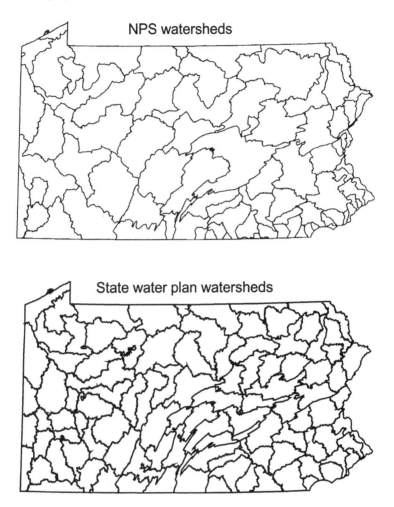

Figure 5.1. The 85 watersheds from the NPS study (above) and the primary set of watersheds, based on the state water plan, for which landscape pattern variables were calculated.

nitrogen deposition can be attributed to local land use activity and since natural "background" sources are negligible, total nitrogen loading was not adjusted for atmospheric deposition. A thematic presentation of total nitrogen loading is seen in Figure 5.4, along with the physiographic provinces of Pennsylvania, where the major provinces are labeled.

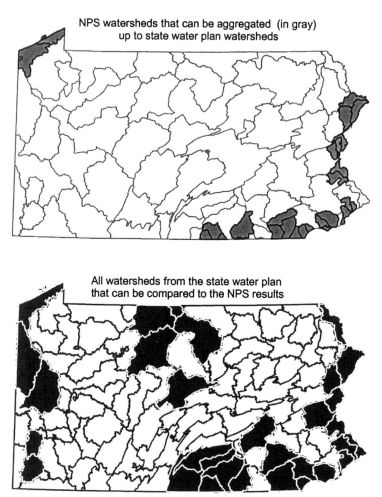

Figure 5.2. Matching of watershed delineations among the Non-Point Source **(NPS)** study and the State Water Plan.

2.2   Pollution Potential Index

Pennsylvania watersheds, as delineated by the state water plan, were ranked via GIS modeling by Hamlett, *et al.* (1992) with respect to their non-point source pollution potential. Various statewide data layers (coverages) were analyzed to produce four different indexes; a runoff index (RI), a chemical use index (CUI), a sediment production index (SPI) and an animal loading index (ALI). An overall pollution potential

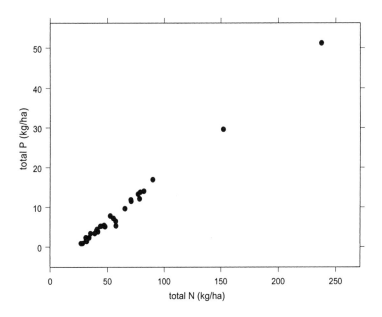

Figure 5.3. Total Phosphorus vs Total Nitrogen (kg/ha).

index (PPI) was then computed for each watershed by:

$$\text{PPI}_i = W_1 \times (\text{RI}_i) + W_2 \times (\text{SPI}_i) + W_3 \times (\text{ALI}_i) + W_4 \times (\text{CUI}_i) \quad (5.1)$$

for the ith watershed, where W_1 to W_4 are weights assigned to the four indices. The results represent per-acre average values. Petersen, *et al.* (1991) show results for an equally weighted version of Equation 5.1 (W_j = 0.25 for $j = 1 \cdots 4$) and a weighted version where the chemical use index is weighted downward to $W_4 = 0.10$ and the remaining input indexes were equally weighted at 0.30. Also, since the model depends heavily on land cover types, results were presented for both "agricultural land" and "all land". While the purpose of the initial study by Hamlett, *et al.* (1992) was to evaluate "agricultural" pollution potential, the purpose of the study being reported in this chapter is to evaluate overall pollution potential. Therefore, we are fortunate that results were also presented by Petersen, *et al.* (1991) for "all lands."

Using the pollution potential for "equally weighted all lands", the resulting watershed ranking is presented thematically in Figure 5.5. The gray scale is the inverse of that presented by Petersen, *et al.* (1991) so

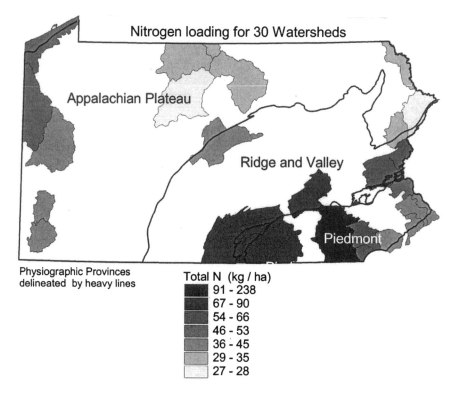

Figure 5.4. Nitrogen loading in kilograms per hectare for 30 watersheds.

that higher pollution potential is represented by darker shading. The watersheds are stratified geographically in Figure 5.5 by aggregating physiographic sections, which are nested within physiographic provinces, in order to form more homogeneous areas with respect to PPI ranks.

The original state water plan delineation, for which PPI values were obtained, consists of 104 watersheds; however, the delineation used for obtaining landscape measurements consists of 102 watersheds resulting from a more spatially accurate aggregation of smaller watersheds that were in turn originally digitized by the USGS. Two of the USGS-source watersheds each consist of two state water plan watersheds; therefore, out of the resulting 102 USGS-source watersheds, two of them did not have direct PPI assessments. For this reason, analysis was limited to 100 of the USGS-source watersheds for which both PPI values and landscape measurements were available. The two "missing" watersheds are indicated by diagonal hatching in Figure 5.5.

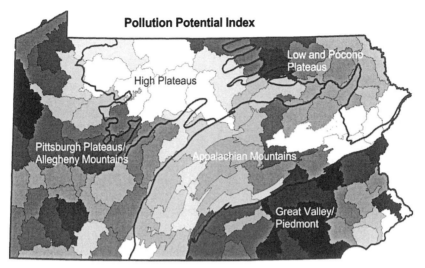

Pollution Potential Index

High Plateaus

Low and Pocono Plateaus

Pittsburgh Plateaus/ Allegheny Mountains

Appalachian Mountains

Great Valley/ Piedmont

Physiographic stratification indicated by heavy black lines

Figure 5.5. Thematic presentation of Pollution Potential Ranking. Increasing shading intensity reflects increasing pollution potential, represented as the inverse rank provided by Petersen, *et al.* (1991) except for two watersheds with missing PPI data, as indicated by diagonal hatching.

3. Selecting an Initial Set of Landscape Pattern Variables

For purposes of regression modeling, the objective is to choose a set of predictor variables that show little to no correlation among themselves in order to avoid multicollinearity. Therefore, an approximately orthogonal subset of spatial pattern variables was obtained by applying principal components analysis (PCA) to the full set of pattern variables in Table 2.1, including the conditional entropy profile parameter values A, B and C. The marginal land cover proportions were excluded from this data reduction exercise because it was desired to include these proportions in the set of potential predictors. Since this set of variables consists of differing measurement units, eigen analysis was performed on the correlation matrix. Results for the 30 watersheds that shared both landscape and nitrogen loading measurements are presented here. When the PCA of the pattern variables was re-applied to all 102 watersheds, the results were essentially the same and are therefore not reproduced here.

As seen in Figure 5.6, the first four components explained over 90% of the variability in the original multivariate data set. Correlations between the original variables and the principal components, which are simply

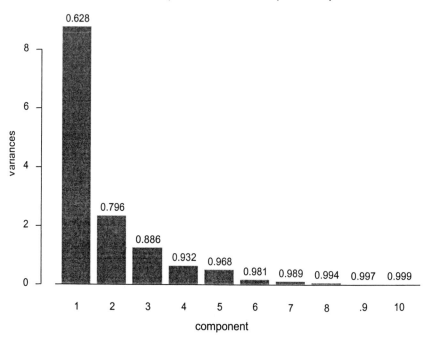

Figure 5.6. Variance contributed by the first ten principal components; cumulative variance is labeled above each bar.

the eigenvector elements (loadings) multiplied by the square root of the corresponding eigenvalue (Stiteler, 1979), are reported in Table 5.1.

The first component is very highly correlated with those landscape matrices that are themselves highly correlated with the marginal land cover proportions. This component reveals the contrast between watersheds that tend towards being more fragmented and more evenly distributed with smaller patches (positive loadings) and those with a high degree of patch coherence (negative loadings). Although many of the original variables could be chosen for representing the first component, contagion (CONTAG) was chosen because it has the highest correlation (in magnitude) in Table 5.1 and is a very familiar measurement in landscape ecology.

The second component is mostly correlated with the conditional entropy profile parameter estimates A and B (note that C is highly corre-

Table 5.1. Correlations between the original variables and the first five principal components.

variable	comp 1	comp 2	comp 3	comp 4	comp 5
PD	0.94	-0.18	0.20	0.07	-0.09
MPS	-0.93	0.22	-0.14	-0.03	0.14
PSCV	-0.78	0.02	0.53	0.16	-0.24
ED	0.94	-0.24	0.13	0.03	0.12
LSI	0.23	0.69	0.07	0.62	0.26
AWMSI	-0.84	-0.09	0.46	0.18	-0.12
DLFD	0.6	-0.04	0.63	-0.24	0.39
AWMPFD	-0.93	-0.01	0.31	-0.1	-0.04
SHEI	0.96	0.19	0.10	-0.07	-0.10
IJI	0.87	0.15	0.15	0.07	-0.38
CONTAG	-0.99	-0.02	-0.10	0.01	0.05
A	0.00	0.93	-0.20	-0.15	-0.11
B	0.26	-0.84	-0.3	0.3	-0.01
C	0.94	0.29	0.01	-0.02	-0.03

lated with component 1, as expected). This component contrasts high values of A, and secondarily the landscape shape index (LSI), as reflected by positive loadings, with high values of B, as reflected by negative loadings.

The third component is most highly correlated with the fractal dimension characterization of patch shape (DLFD) and secondarily with the patch size CV (PSCV). The fourth component is dominated by the landscape shape index.

In view of the results of principal components analysis, the spatial pattern variables that were included in the set of potential regressors were patch size coefficient of variation (PSCV), landscape shape index (LSI), fractal dimension (DLFD), contagion (CONTAG) and the conditional entropy profile values A and B. Note that the remaining conditional entropy profile value C could have also been chosen in place of contagion, as per the very high correlation with component 1 in Table 5.1 and the highly linear relationship between contagion and C in Figure 4.3. Finally, the proportions of annual herbaceous land (ANN.HERB), total herbaceous land (TOT.HERB), which is the sum of annual and perennial herbaceous land, and total forest land (TOT.FOREST), which is the sum of broadleaf, conifer and mixed forest lands, were added to the set of potential regressors.

Relationships among the final set of potential landscape predictors for the sample of 30 watersheds containing both landscape and nitrogen loading measurements are seen in Figure 5.7, where total nitrogen is also included as a log transform (logN) for reasons discussed later. One expects the proportion of annual herbaceous land to be a very strong, if not dominant, predictor of total nitrogen loading since it consists mostly of cropland. Indeed, agriculture was determined to be a main source of nitrogen loading in the initial study (Nizeyimana, *et al.* 1997). This expectation is borne out by the scatter plots involving logN in Figure 5.7. It is also interesting to note the negative correlation between logN and TOT.FOREST, which probably reflects the negative correlations between total forest cover and the two herbaceous categories.

Relationships among the variables for all of the 102 watersheds are presented in Figure 5.8 along with the inverse of the Pollution Potential Index rank (PPI.INV).

The three land cover proportions plotted in Figures 5.7 and 5.8 are highly inter-correlated, as expected, and some redundancy exists between PSCV and CONTAG as well as between the values of A and B; however, it is desired to include all of these variables in the initial set of landscape measurements in order to see which may be chosen over others as part of stepwise model building, as discussed in Section 4.

4.   Linear Models for Relating Water Pollution Loading to Landscape Variables

Stepwise regression was applied separately for each response variable, total nitrogen loading and the pollution potential index, in order to find an "optimal" set of regressors from among the candidate regressors in Figures 5.7 and 5.8. Using S-Plus© software, the criterion for choosing the best set of predictors was a modification of Mallow's C_p statistic (Mallows, 1973), known as the Akaike Information Criterion (AIC) (Akaike, 1974). The AIC is related to the C_p statistic by the relation

$$\text{AIC} = \text{MSE}(C_p + n),$$

for n observations and where MSE equals the mean squared error of the current model before adding or deleting a term to yield an updated p-parameter model (MathSoft, Inc., 1997, p. 132). An explicit expression for AIC is

$$\text{AIC} = \text{RSS}(p) + \text{MSE} * 2 * p, \tag{5.2}$$

where $\text{RSS}(p)$ is the residual sum of squares from the updated model containing p terms (k predictors plus the intercept) and MSE is the mean squared error from the original model prior to deleting or adding a term.

The automated stepwise selection procedure works by choosing the set of predictors that minimizes the AIC statistic. Critical F values for

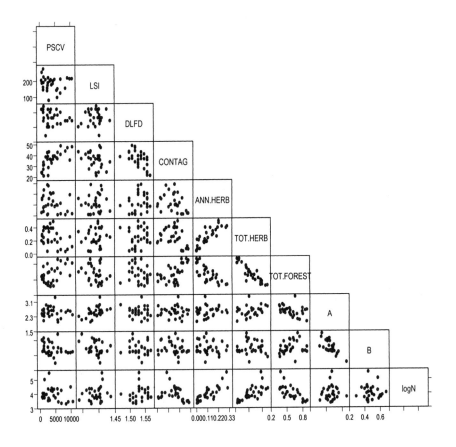

Figure 5.7. Pairwise scatterplots of the final set of potential predictor variables (regressors) along with the natural logarithm of total nitrogen per hectare. The spatial pattern variables are as follows: PSCV = Patch Size Coefficient of Variation, LSI = Landscape Shape Index, DLFD = Double Log Fractal Dimension, CONTAG = contagion and conditional entropy profile parameter estimates (A, B). Marginal land cover proportions are ANN.HERB = annual herbaceous, TOT.FOREST = total forest and TOT.HERB = total herbaceous.

deciding whether or not to include or remove predictor variables was set at 2, thereby favoring larger sets of predictor variables.

Models were checked using standard diagnostic graphics. In addition, partial residual plots were obtained for each regressor in a model (Montgomery and Peck 1982). Partial residual plots display the relationship between the response y and the regressor x_j after the effect of all other

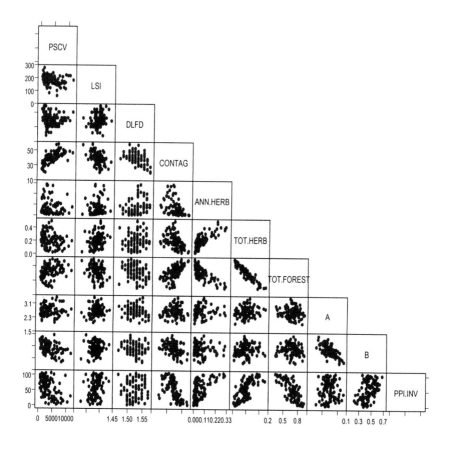

Figure 5.8. Pairwise scatterplots of the set of potential predictor variables (regressors) along with the inverse of the pollution potential index (PPI.INV), as presented in Figure5.5 for 102 watersheds. The landscape variables are explained in Figure 5.7.

regressors $x_i (i \neq j)$ have been removed. Along with providing a check for outliers and inhomogeneity of variance, these plots can also suggest possible linearizing data transformations.

4.1   Predicting In-Stream Nitrogen Loading

Initial analysis was performed using total nitrogen (kg/ha) as the response variable; however, the resulting model was excessively influenced by two watersheds from the Piedmont physiographic province (see Figure 5.4). A natural log transform substantially reduced the domineering influence of these two watersheds and yielded other diagnostics that were much better; therefore, all analyses proceeded with the log of total nitrogen (logN) as the response variable.

The graphical relationship of logN with the potential predictor variables is seen in Figure 5.7. Although these scatter plots show a fairly strong linear relationship between logN and the marginal land cover properties, other potential predictors may also explain a significant portion of the variability in observed logN. In fact, bivariate scatter diagrams can be misleading in the multiple regression context, as pointed out by Montgomery and Peck (1982, p. 122), who cite Daniel and Wood (1980).

Preliminary analysis showed that when all 30 of the NPS watersheds were included, the only variable retained by the stepwise selection procedure was the proportion of annual herbaceous land (ANN.HERB); however, when separate analyses were performed for each major physiographic province, very different results were obtained. For the 12 watersheds of the Appalachian Plateaus, all but the landscape shape index were retained. Since the Ridge and Valley had only 7 NPS watersheds, they were combined with the Piedmont watersheds, which have similar forest fragmentation patterns. For the 18 watersheds of the combined Piedmont / Ridge and Valley Province group, annual herbaceous land was retained along with the fractal dimension (DLFD) and both the A and B values of the conditional entropy profiles.

In light of the Province-specific dependence of the models and with a view toward increasing the residual degrees of freedom, the analysis was continued by combining all 30 watersheds, but including an indicator variable in the model to specify membership in a physiographic region. The indicator variable, which was forced to be retained by the model, was coded with "1" if the corresponding watershed was from the Piedmont/Ridge and Valley group, and a "0" otherwise. The resulting parameter estimate revealed the increase (or decrease) in total nitrogen loading as one moves from the Appalachian Plateaus to the Piedmont/Ridge and Valley group.

Coefficient estimates for the model that minimized the AIC statistic (AIC = 2.84) are reported in Table 5.2, where the dummy variable indicating the effect of province group is labeled as PIED.RV.

Diagnostic plots for the fitted model given in Table 5.2 are reported in Figure 5.9. The residuals show no obvious patterns other than a slight preponderance of negative residuals and a tendency for large magnitude residuals to be positive. The fitted values show a strong linear relationship with the observed values. Also, the spread of residuals is tighter than the spread of fitted values, as seen in the lower middle plot in Fig-

Table 5.2. Coefficients and corresponding statistics from regressing the log of total Nitrogen/ha against quantitative landscape variables and an indicator variable for specifying membership in a physiographic province group. Mean squared error (24 d.f.) = 0.075 and multiple $R^2 = 0.74$.

Regressor*	Value	t value	p value
Intercept	12.781	2.58	0.017
Pied.RV	0.418	2.14	0.043
LSI	0.003	2.31	0.030
DLFD	−5.893	−1.85	0.077
ANN.HERB	3.244	3.99	0.001
A	−0.410	−1.81	0.084

*Pied.RV indicates membership in the Piedmont/Ridge and Valley group of physiographic provinces, as opposed to membership in the Appalachian Plateau Province.
LSI = Landscape Shape Index, DLFD = Double LogFractal Dimension, ANN.HERB = proportion of Annual Herbaceous land, and
A = conditional entropy profile depth.

ure 5.9, which is like a graphical equivalent to the multiple R^2 statistic. The Q/Q plot in the lower left side of Figure 5.9 reveals somewhat heavy tails in the distribution of residuals; however, none of these observations are excessively influential according to Cook's Distance, as seen in the lower right of Figure 5.9. Generally, a Cook's Distance of 1 or greater is considered to reveal an overly influential observation (Montgomery and Peck 1982; Neter, Wasserman and Kutner, 1985); but this critical value is much greater than the worst case (0.3) reported in Figure 5.9. Consequently, the diagnostics indicate an acceptable model.

The partial residual plots for each quantitative predictor in Table 5.2 appear in Figure 5.10. The lines of fit in Figure 5.10 have slopes equal to the parameter estimates in Table 5.2. The plots in Figure 5.10 indicate a linear trend for each predictor, especially for annual herbaceous land (ANN.HERB), and no data transformations appear to be necessary.

4.2   Predicting a Pollution Potential Index

For the purpose of regression modeling, the five geographic strata that appear in Figure 5.5 are represented by four indicator variables that are explained in Table 5.3. These indicators were forced to be retained by the model selection protocol in order to factor out physiographic effects and reduce effects of possible spatial autocorrelation. The resulting parameter estimates reveal the increase (or decrease) in average PPI rank as one moves from the "Pittsburgh Plateaus/Allegheny Mountains" group

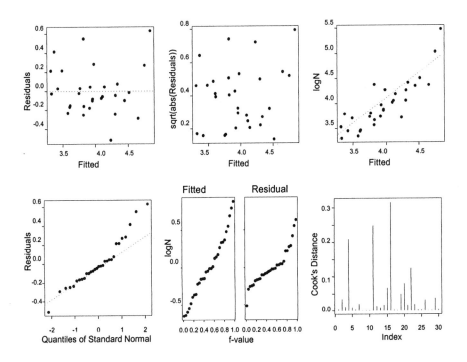

Figure 5.9. Diagnostic plots corresponding to the regression model defined in Table 5.2.

to the group being represented by the respective indicator variable. The model fit with the minimum AIC statistic is given in Table 5.3.

Diagnostic plots for the fitted model given in Table 5.3 are reported in Figure 5.11. The residuals appear randomly scattered with no obvious trend and the fitted values show a strong linear relationship with the observed values. Also, the spread of residuals is tighter than the spread of fitted values. The Q/Q plot in the lower left side of Figure 5.11 reveals that the residuals are closely approximated by the normal distribution. None of these observations are excessively influential according to Cook's Distance, as seen in the lower right of Figure 5.11. Consequently, these diagnostics reveal a very acceptable model.

The partial residual plots for each quantitative predictor in Table 5.3 appear in Figure 5.12. The lines of fit in Figure 5.12 have slopes equal to the parameter estimates in Table 5.3. The plots in Figure 5.12 indicate

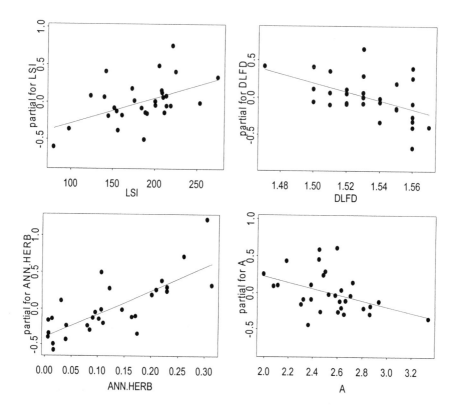

Figure 5.10. Partial residual plots for the predictors listed in Table 5.2. Slopes of the fitted lines equal the parameter estimates in Table 5.2.

a linear trend for each predictor, and no data transformations appear to be necessary.

5.　Interpretation

The chosen model for relating total nitrogen loading (kg/ha) to landscape characteristics within Pennsylvania watersheds that are delineated based on the state water plan is as follows:

$$\ln(N) \;=\; 12.78 + 0.42(\text{Pied.RV}) + 3.24(\text{ANN.HERB})$$
$$+0.0034(\text{LSI}) - 5.89(\text{DLFD}) - 0.41(A), \qquad (5.3)$$

where the associated statistics for the parameter estimates based on a sample of 30 watersheds, and an explanation of the variable labels

Table 5.3. Linear coefficients from regressing the PPI rank against quantitative landscape variables and physiographic indicator variables. Mean squared error = 245.55 (90 d.f.) and multiple $R^2 = 0.76$.

Regressor*	Coefficient	t value	p value
Intercept	445.228	1.79	0.076
APP. MOUNTAIN	−6.275	−1.09	0.279
PIED. and GR. VALLLEY	15.720	2.44	0.017
LOW and POCONO	−2.352	−0.35	0.725
HIGH PLATEAUS	−12.429	−1.88	0.064
DLFD	−330.221	−2.46	0.016
CONTAG	−0.973	−2.19	0.031
TOT.HERB	118.906	5.01	0.000
A	27.801	2.26	0.026
B	110.380	2.61	0.012

*Labels for the quantitative variables are explained in Figure 5.8
APP. MOUNTIAN = Appalachian Mountain Section
PIED. and GR.VALLEY = the Piedmont Plateau and Great Valley Section
LOW and POCONO = Glaciated Low and Pocono Plateau Sections
HIGH PLATEAUS = High and Mountainous High Plateau Sections

are found in Table 5.2. The associated variance σ^2 is estimated by MSE=0.075, although this might be a slight underestimate due to some spatial autocorrelation in the Appalachian Plateaus.

As expected, the dominant regressor is the proportion of annual herbaceous land, which is mainly cropland. Given the proportion of annual herbaceous land and different physiography, landscape *pattern* strengthened the explanation of nutrient loading variability among these Pennsylvania watersheds, as measured through total nitrogen loading. The landscape shape index (LSI), the fractal dimension estimate (DLFD) and the estimate of conditional entropy profile depth (A) were all retained by the stepwise selection procedure which aims to minimize the AIC statistic across all possible regressions.

The numerically small, but statistically significant, coefficients of the landscape shape index indicates that nitrogen loading may be expected to increase as the landscape becomes more fragmented, resulting in more edges.

A negative relation to the value A is not readily interpretable; however, it is noteworthy that this predictor and LSI were both retained by the stepwise selection procedure whether the physiography indicator variables were designed to differentiate among the 3 major provinces (results not shown here), the 2 province groups (Appalachian Plateaus vs.

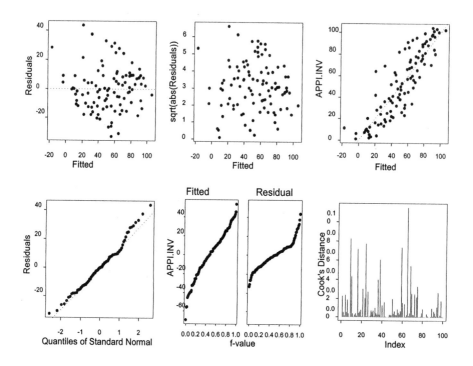

Figure 5.11. Diagnostic plots corresponding to the regression model defined in Table 5.3.

Piedmont/Ridge and Valley) or the 3 groups that consisted of the Piedmont/Ridge and Valley and 2 sub-areas of the Appalachian Plateaus.

The chosen model for relating the pollution potential index (PPI) rank to landscape characteristics within Pennsylvania watersheds that are delineated based on the state water plan is as follows:

$$
\begin{aligned}
\text{PPI rank} \;=\; & 445.2 - 6.3(\text{APP. MOUNTIAN}) \\
& +15.72(\text{PIED. and GR. VALLLEY}) \\
& -2.35(\text{LOW and POCONO}) - 12.43(\text{HIGH PLATEAUS}) \\
& +118.9(\text{TOT.HERB}) - 330.2(\text{DLFD}) - 1.0(\text{CONTAG}) \\
& +27.8(A) + 110.4(B), \qquad\qquad\qquad\qquad (5.4)
\end{aligned}
$$

where an explanation of the variable labels is found in Figure 5.8 and Table 5.3.

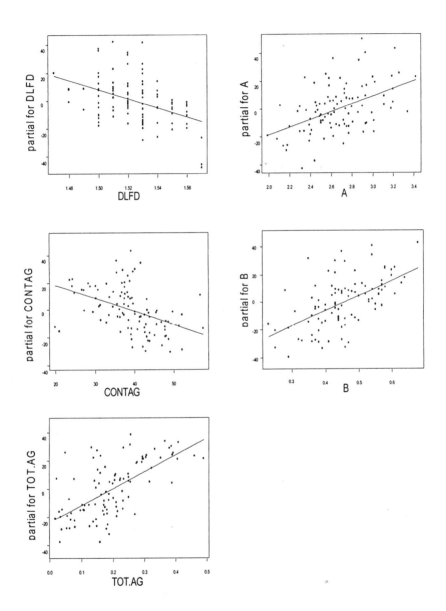

Figure 5.12. Partial residual plots for the quantitative predictors listed in Table 5.3. Slopes of the fitted lines equal the linear coefficients in Table 5.3.

The dominant regressor is the proportion of total herbaceous land; however, our results show that given the proportion of total herbaceous land and adjusting for physiography, landscape pattern strengthens the prediction of surface water pollution potential across these Pennsylvania watersheds.

The negative relation to fractal dimension is consistent with the nitrogen loading results. A negative relation makes sense because when landscape patches are left to natural forces, they tend towards having more irregular outlines, which is reflected by an increasing fractal dimension (or perimeter/area scaling exponent) (Johnson, Tempelman and Patil, 1995), while patches that are created and maintained by humans tend to have straight edges, especially with cropland that is in turn largely responsible for nutrient loading. As the average landscape patch tends towards having a straighter edge, the fractal dimension metric decreases. A negative relation to contagion is likely due to the highest levels of contagion being associated with mostly forested watersheds. Although both conditional entropy profile variables A and B are retained by the stepwise protocol, a mechanistic explanation of the relation between PPI is not necessarily clear.

As an exploratory exercise, nine watersheds were chosen to include the top three, middle three and lowest three nitrogen loading values, and this was repeated for the PPI values. Their corresponding conditional entropy profiles appear in Figures 5.13 and 5.14. For both nitrogen loading and the PPI, the three least polluted watersheds are clearly separate from the others which, in turn, are essentially grouped together. These three watersheds with the lowest pollution potential are mostly forested watersheds from the High Plateaus or Poconos and consistently reveal lower profiles that are "intrinsically less fragmented" than the other six profiles. Although these profiles do not reveal apparently large differences in A and B values, the model for predicting nitrogen loading benefitted from including A and the ability to predict pollution potential was strengthened by including both A and B.

In summary, the best landscape-level predictor of water pollution for these Pennsylvania watersheds is the marginal land cover proportions. A majority of nitrogen loading variability was explained by the proportion of annual herbaceous land, which is mostly row crops. Meanwhile, variability of the pollution potential index was largely explained by total herbaceous land, which includes annual and perennial herbaceous land. This finding agrees with results by Roth, Allan and Erickson (1996), who found that stream biotic integrity was significantly correlated with the proportion of agricultural land throughout a whole watershed. These authors further concluded that stream conditions are primarily deter-

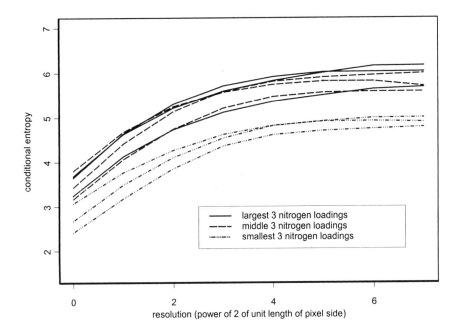

Figure 5.13. Conditional entropy profiles for watersheds containing the top 3, middle 3 and bottom 3 nitrogen loadings.

mined by regional land use, overwhelming the ability of local riparian vegetation to support high quality habitat. Hunsaker and Levine (1995) also determined that nitrogen, phosphorous and conductivity were all primarily dictated by land use proportions and they further cite other studies that lead to similar findings. This is all quite encouraging because once a reliable land cover map is in place, the marginal land cover proportions are readily available; therefore, without any further information, one can make a fairly strong prediction of surface-water quality within a watershed. Land cover proportions within a watershed may be even more informative if interaction with season is considered (Mehaffey, *et al.*, 2003).

We have shown, however, that inclusion of spatial *pattern* variables can significantly strengthen the predictability of pollution loading within these Pennsylvania watersheds. Furthermore, some aspects of the multi-resolution conditional entropy profiles were consistently retained by an objective variable selection protocol.

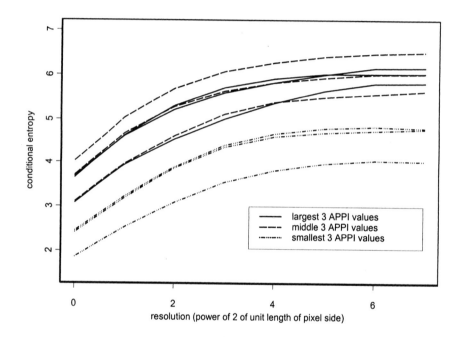

Figure 5.14. Conditional entropy profiles for watersheds containing the top 3, middle 3 and bottom 3 PPI values.

Chapter 6

PREDICTABILITY OF BIRD COMMUNITY-BASED ECOLOGICAL INTEGRITY USING LANDSCAPE VARIABLES

1. Introduction

In Chapter 4, watershed clusters that were based on landscape measurements alone were categorized with respect to "apparent" ecological condition based on the type and degree of forest fragmentation. This chapter, based on Johnson, *et al.* (2003), serves to provide an independent categorization of the watershed-delineated landscapes using only bird community data for identifying categories of ecological integrity. Therefore, a more objective basis is obtained for determining how well the landscape measurements alone can discriminate among watersheds with respect to ecosystem condition.

The ecological data source is the Pennsylvania breeding bird atlas (Brauning and Gill, 1983–1989), which is the result of a five-year field survey performed by trained volunteers where the presence of breeding evidence is recorded for bird species in each of approximately 5000 blocks covering the entire state. Each block is one sixth of a USGS 7.5 minute quadrangle. Encountered species were assigned one of four different levels of strength of evidence for breeding. For this current study, any species listed under one of the three strongest evidence categories ("possible", "probable" and "confirmed") was treated as "present". Atlas blocks were assigned to a given watershed if the block's center was within the watershed boundary, as depicted in Figure 6.1.

Given the set of atlas blocks for a watershed, where each block contains a species list, the object is to use these data to assess ecological condition for the watershed. Indices of species richness and diversity are insufficiently sensitive to changes in community composition to be useful for our purpose. This is especially true for geographic areas as large

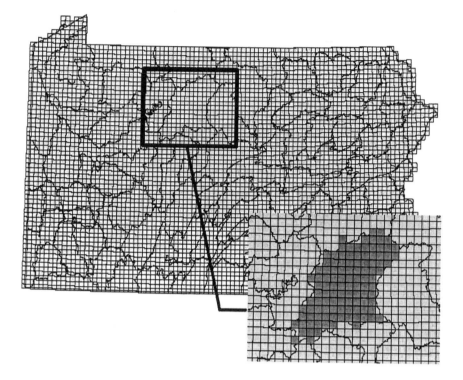

Figure 6.1. Breeding Bird Atlas blocks (small squares) and State Water Plan-based watersheds (large, irregular polygons).

as these watersheds. However, if a species list is converted to response guilds that represent structural, compositional and functional components of an ecosystem, then guild composition can effectively assess ecological condition (Brooks, O'Connell, Wardrop and Jackson, 1998). First we describe some guilding schemes that aim to provide a meaningful ecological assessment at the watershed level.

2. Assessing Ecological Integrity using Songbird Community Composition

2.1   Background

A landscape-level indicator of ecological condition was developed by O'Connell, Jackson and Brooks (1998ab) for the Mid-Atlantic Highlands Assessment (MAHA) area that is depicted in Figure 6.2. With the motivation of characterizing ecological condition in the same manner as an Index of Biotic Integrity (Bradford, *et al.*, 1998; Karr, 1991, 1993),

Figure 6.2. The Mid-Atlantic Highlands Assessment Area, shaded in gray, within Region 3 of the U.S. EPA.

an indicator was developed based on the breeding songbird community composition that was obtained through representative field sampling. A bird community with high integrity would be dominated by guilds that depend on native system attributes. The indicator incorporates aspects of ecosystem structure (e.g., vegetation), function (e.g., energy flow) and composition (e.g., demographics). Since the landscape-scale native vegetation matrix for Pennsylvania is temperate forest, then specialist guilds such as obligate tree-canopy nesters indicate high integrity, while generalist guilds such as shrub nesters indicate low integrity.

O'Connell, Jackson and Brooks (1998ab) initially evaluated their approach using 34 reference sites. First, these sites were grouped into high, medium and low integrity categories based on the bird community index (BCI). This grouping was then compared to an independent *a priori* classification of the same sites as pristine, moderate and disturbed. The *a priori* classification incorporated factors such as sediment deposition, soil properties, plant and amphibian community structure, wildlife community habitat and general landscape context. Upon observing satisfactory correspondence between the two independent ranking approaches, a blocked random sample was selected consisting of 126 sites covering the entire MAHA area. Sampling was based on the national Environmental Monitoring and Assessment Program (EMAP) sampling grid (Overton, 1990). Field sampling details are found in O'Connell, Jackson and Brooks (1998a).

Each site was represented by a list of guilds, where each guild was represented by the proportion of species observed at the site that belong to the respective guild. Since a species could belong to several guilds, the list of proportions at a site does not necessarily sum to one. The sites were clustered in guild space using the complete linkage squared Euclidean distance protocol. Next, an analysis of variance was done for each guild separately, using cluster membership as factor levels and species proportions as the response. Tukey's simultaneous confidence intervals ($\alpha = 0.05$) were then used to determine the statistically separable clusters of sites for each guild. This was performed iteratively, combining clusters that were not significantly different from each other. If a guild did not yield at least two statistically distinguishable clusters of sites, then the guild was considered non-informative and therefore eliminated. In this manner, sixteen guilds were identified, where each guild yielded 2 to 5 statistically separable clusters of sites.

Ranking scores based on the species proportions were then assigned to these clusters for each guild, whereby if a guild was a specialist, the ranks increased as the species proportions increased and if a guild was a generalist, the ranks decreased as the species proportions increased. Therefore, a set of ranks was obtained for each guild, where each rank corresponded to a range of species proportions, as presented in Table 6.1. Fractional ranks occur from averaging ranks of statistically indistinguishable clusters. In cases where clusters failed to be statistically distinguishable, yet only slightly overlapped, then separate groups may have still been identified. Therefore, the final ranking scheme reported in Table 6.1 is actually the result of interactive statistical analysis and expert judgment.

Table 6.1. Biotic integrity ranks for 16 songbird guilds in the Mid-Atlantic Highlands Assessment area.

guild type	guild	proportion	rank
structural	forest birds	0.000–0.280	4.5
		0.281–1.000	2.5
	interior forest birds	0.000–0.010	1
		0.011–0.080	1.5
		0.081–0.260	3
		0.261–0.430	4
		0.431–1.000	5
	forest ground–nesters	0	1
		0.001–0.020	1.5
		0.021–0.160	3
		0.161–0.240	4.5
		0.241–1.000	5
	open ground–nesters	0.000–0.020	1
		0.021–0.110	2.5
		0.111–1.000	5
	shrub–nesters	0.000–0.210	4
		0.211–0.330	1.5
		0.331–1.000	1
	tree canopy–nesters	0.000–0.280	1.5
		0.281–0.320	2
		0.321–1.000	4.5
functional	bark–probing insectivores	0.000–0.060	1.5
		0.061–0.110	3
		0.111–0.170	4
		0.171–1.000	5
	ground–gleaning insect.	0.000–0.050	1.5
		0.051–0.070	2
		0.071–0.140	4.5
		0.141–1.000	5
	tree canopy insectivores	0.000–0.030	1.5
		0.031–0.050	2
		0.051–0.120	3
		0.121–0.200	4.5
		0.201–1.000	5
	shrub–gleaning insect.	0.000–0.140	1.5
		0.141–0.230	2.5
		0.231–1.000	5

(continued ...)

continuation of Table 6.1

guild type	guild	proportion	rank
	omnivores	0.000–0.290	5
		0.291–0.410	4
		0.411–0.480	3
		0.481–0.580	1
		0.581–1.000	2
compositional	nest predator/brood parasite	0.000–0.100	5
		0.101–0.150	3.5
		0.151–0.180	2
		0.181–1.000	1
	exotic species	0	5
		0.001–0.020	4.5
		0.021–0.050	3
		0.051–0.110	2
		0.111–1.000	1
	residents	0.000–0.260	5
		0.261–0.390	3.5
		0.391–0.570	2
		0.571–1.000	1
	temperate migrants	0.000–0.210	4
		0.211–0.300	2
		0.301–1.000	1
	single–brooded	0.000–0.410	1.5
		0.411–0.450	2
		0.451–0.610	3
		0.611–0.730	4
		0.731–1.000	5

Each site was then assigned a "bird community index" (BCI) score by summing the ranks assigned to each guild within the respective site. The logic here is that the BCI reflects biotic integrity because the BCI increases as the ratio of specialists to generalists increases. One can also break down the overall BCI into functional, compositional and structural scores, which is a valuable feature of this approach (O'Connell, Jackson and Brooks, 1998a).

The 5 distinct clusters of sites were ranked according to the relative proportions of specialists and generalist guilds at the sites within each cluster. This ranking scheme allowed the placement of 5 clusters into 4 distinct categories of BCI scores, labeled as "low", "medium","high" and "highest". The range of site BCI scores that correspond to each category is presented in Table 6.2. More detailed analysis of individual species proportions in each guild allowed separation of the low-integrity group into "low-agricultural" and "low-urban" categories.

Table 6.2. BCI scores corresponding to qualitative categories of ecological integrity.

highest integrity:	60.1–77.0
high integrity:	52.1–60.0
medium integrity:	40.1–52.0
low integrity:	20.0–40.0

2.2 Application to Breeding Bird Data

Using the songbird species and final set of 16 guilds from O'Connell, Jackson and Brooks (1998a), presence/absence data were obtained from the Breeding Bird Atlas and an index of biological integrity was constructed as described below in Sections 2.3 and 2.4. The proportion ranges in Table 6.1 were used for assigning the species proportions within a guild to a ranking score, and the values in Table 6.2 were used for mapping the BCI values to an integrity category.

Since the results in Tables 6.1 and 6.2 are based on a probability sample for the entire MAHA area, we only applied our protocol to those Pennsylvania watersheds that intersected the MAHA region. The native system conditions are expected to be similar at the landscape scale throughout the MAHA area because this area lies within the Appalachian Plateaus and Ridge and Valley physiographic provinces and is affected by common climatic regimes.

Note that results from O'Connell, Jackson and Brooks (1998a) are from sampling sites that were 79 ha, whereas an atlas block is approximately 2800 ha and different data collection methods were used in the two investigations. Also, some known sampling bias occurs in both the Breeding Bird Atlas (Brauning, 1992) and the study by O'Connell, Jackson and Brooks (1998a). While these limitations should be noted, we are nevertheless presented with a valuable spatially synoptic database that we are analyzing in a logical way to extract ecological integrity assessments for whole watersheds.

2.3 Block-Level BCI Values

For each of $w = 1, \cdots, W$ watersheds, containing $i = 1, \cdots, N_w$ breeding bird atlas blocks, let S_{wi} be the number of species in the ith block of watershed w and let S_{wij} be the number of species in the jth guild of the ith block of watershed w.

For each block, a guild profile is obtained by computing the proportion of overall species in the block that belong to the various guilds. Specifically, for the ith block in the wth watershed, a species proportion is computed for the jth guild by

$$P_{wij} = \frac{S_{wij}}{S_{wi}}, \quad j = 1, \cdots, G.$$

As explained above, $\sum_{j=1}^{G} P_{wij}$ is generally not equal to 1 since a given species can belong to several guilds.

Watershed w is then represented by a matrix of N_w rows (blocks) and G columns (guilds),

$$\mathbf{P}_w = \begin{bmatrix} P_{11} & \cdots & \cdots & \cdots & P_{1G} \\ \vdots & \ddots & & & \vdots \\ \vdots & & P_{wij} & & \vdots \\ \vdots & & & \ddots & \vdots \\ P_{N_w 1} & \cdots & \cdots & \cdots & P_{N_w G} \end{bmatrix}.$$

Each entry P_{wij} is then converted to a rank score using the columns labeled "proportions" in Table 6.1. This yields the following matrix of rank scores:

$$\mathbf{R}_w = \begin{bmatrix} R_{11} & \cdots & \cdots & \cdots & R_{1G} \\ \vdots & \ddots & & & \vdots \\ \vdots & & R_{ij} & & \vdots \\ \vdots & & & \ddots & \vdots \\ R_{N_w 1} & \cdots & \cdots & \cdots & R_{N_w G} \end{bmatrix}.$$

Each row of the $\mathbf{R}_w$ matrix is then summed to yield a "Bird Community Index" (BCI) for each block (row), where the ecological integrity increases with increasing BCI. The resulting BCI value for each block is converted to one of four categories of ecological integrity–"low", "medium", "high" and "highest"– based on BCI cutoff values in Table 6.2.

Results for all of the Pennsylvania watersheds that are in the MAHA region are presented in Figure 6.3. Each watershed can now be described by the proportion of land (proportion of blocks) that is in each of the four categories of ecological integrity.

2.4   Watershed-level Ecological Integrity

The next question is how to summarize the information in Figure 6.3 so that each watershed can be assigned some overall value of ecological integrity. One can rank the watersheds according to the proportion of

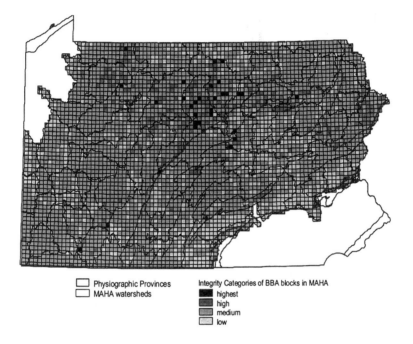

Figure 6.3. Breeding Bird Atlas blocks coded according to categories of ecological integrity with respect to the songbird community, overlaid with Pennsylvania "State Water Plan-based" watersheds that are in the MAHA region and the major physiographic provinces of Pennsylvania.

land in one of the integrity categories. Ranking by "medium–integrity" land proportions would not be very informative. Ranking by "highest–integrity" land proportions is also not very helpful because only 0.6% of the blocks were in this category.

The watersheds could be ranked based on the proportion of "low-integrity" or "high–integrity" land; however, one must then decide how to group the ranks into sensible categories of integrity. It seems most informative to characterize each watershed based on the full spatial distribution of ecological integrity; therefore, the following approach was taken.

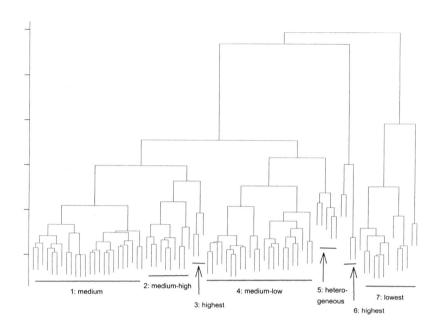

Figure 6.4. Dendrogram of watershed clusters in the Pennsylvania MAHA region obtained from applying the average Euclidean distance agglomerative clustering protocol to the proportions of "low","medium", "high" and "highest" integrity atlas blocks in each watershed.

All of the MAHA watersheds were clustered using the "average Euclidean distance" clustering protocol, where the response variables were the proportions of land in each of the four integrity categories (low, medium, high and highest). Average linkage was used for reasons of robustness and consistency. Further, when Manhattan Distance was evaluated, no differences were seen in the resulting clusters. The clustering was done collectively for all the MAHA–region watersheds, as opposed to separate analysis for the physiographic provinces, because the original study that yielded the P_{wij} and BCI cutoffs (O'Connell, Jackson and Brooks, 1998a) was performed on sampled sites from throughout MAHA.

The resulting cluster dendrogram is seen in Figure 6.4 and the watershed clusters are mapped in Figure 6.5. The general category labels

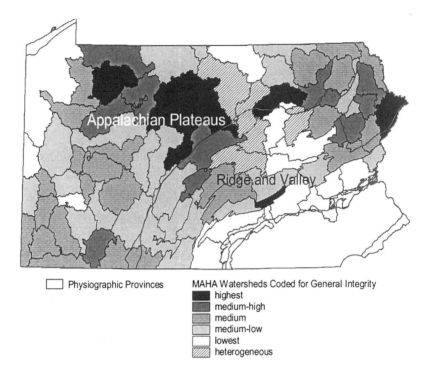

Figure 6.5. Watersheds in the MAHA region coded according to general ecological integrity as determined by cluster membership in Figure 6.4.

for the clusters in Figure 6.4 were based on the distributions of the four integrity categories as summarized in the box plots of Figure 6.6.

All watersheds in cluster 7 have a higher proportion of low-integrity blocks than any watersheds of other clusters. Since cluster 7 also has the lowest distribution of medium– and high–integrity blocks, along with no occurrence of highest–integrity blocks, then this cluster is clearly a "lowest" general integrity group.

At the other extreme, watersheds of clusters 3 and 6 collectively have a higher proportion of high–integrity blocks than any watersheds of other clusters. Since these clusters also yield among the lowest overall proportions of low-integrity blocks, then clusters 3 and 6 are clearly in a "highest" general integrity group.

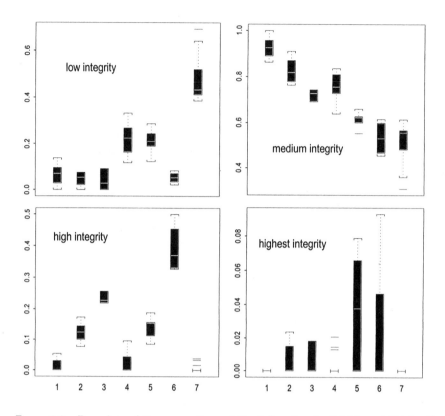

Figure 6.6. Box plots of the proportions of "low","medium", "high" and "highest" integrity land in each of the seven clusters identified and labeled in Figure 6.4. The numbers 1 to 7 appearing along the bottom axes correspond to the labeled clusters in Figure 6.4 as they occur from left to right.

Cluster 1 is clearly a "medium" general integrity group, as these watersheds range from having about 85% to 100% medium–integrity blocks, have among the lowest proportions of both low- and high–integrity blocks and no blocks that are highest–integrity.

Watersheds of clusters 4 and 5 have the second highest proportions of low-integrity blocks. Since cluster 4 watersheds have a relatively high proportion of medium–integrity blocks and also have among the lowest proportions of high–integrity blocks, then cluster 4 appears to be in a "medium–low" general integrity category.

Watersheds of clusters 2 and 5 collectively have the second largest proportions of high–integrity blocks. Since the watersheds of cluster 2 have among the lowest proportions of low-integrity blocks and contain some highest–integrity blocks, yet are predominantly medium–integrity, then cluster 2 appears to be a "medium–high" general integrity category.

Finally, cluster 5 is labeled "heterogeneous" because of the simultaneous high proportion of both low- and high–integrity blocks, with some highest–integrity. Such heterogeneity can be explained by very different land uses in different areas of the watersheds. Allocating this group into the other categories resulted in substantially increasing the variability seen in the box plots of Figure 6.6; therefore, it was decided to keep these five watersheds in a separate group.

The results mapped in Figure 6.5 generally agree with intuitive expectation based on landscape context, with the only apparent anomaly being the one watershed on the northern border (Tioga Creek) that is labeled as "lowest". Based on land cover alone, one may expect a "medium" classification at worst for this particular watershed; however, these results may reflect lower sampling coverage in this area (Brauning, 1992). On the other hand, a small watershed in the Ridge and Valley province (Clarke Creek) was placed into the highest category, in contrast to its neighboring watersheds. This is, however, the appropriate classification since Clarke Creek is the protected drinking water supply for the city of Harrisburg and it is heavily forested.

3. Relating Landscape Attributes to the Songbird-based Assessments of Watershed–wide Ecological Integrity

Next we study whether and how well ecological integrity can be predicted by the landscape variables discussed in Chapter 2 on a watershed-wide basis.

Box plots of an approximately orthogonal subset of the spatial pattern measurements are presented in Figure 6.7 for each of the seven ecological integrity clusters in Figure 6.4. Similarly, boxplots for the three land cover proportions are presented in Figure 6.8. These results indicate that contagion, total forest and total herbaceous land reveal monotonic trends as one moves from the highest to lowest general integrity.

3.1 Linear Regression Modeling

The relationship between ecological integrity and landscape variables was attempted to be quantified by regression modeling, where the response variable was either the proportion of low integrity land, the sum of both high and highest integrity land or the raw BCI scores averaged over the watersheds. For each of the different responses, a stepwise selection protocol was applied, whereby an optimal set of predictors was

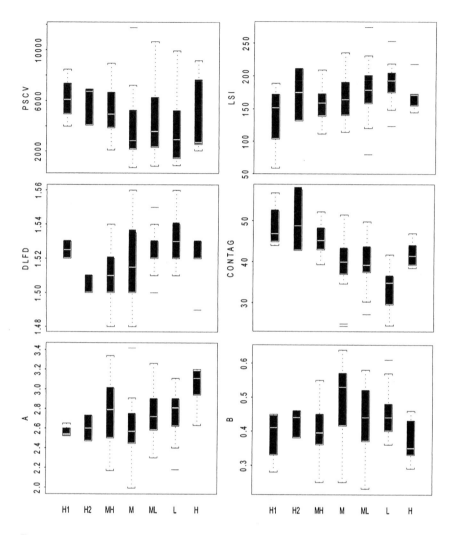

Figure 6.7. Box plots of the landscape pattern variables for each of the seven clusters identified in Figure 6.4. The cluster integrity levels are labeled as follows: H1 and H2 = highest, MH = medium–high, M = medium, ML = medium–low, L = lowest and H = heterogeneous. The horizontal axis shows integrity from highest to lowest, except that the "heterogeneous" category (H) cannot be conveniently located along an integrity gradient.

chosen by minimizing the AIC statistic, as discussed in Chapter 5. A dummy variable was forced to be retained in each model to account for the effect of membership in the Appalachian Plateaus, relative to the Ridge and Valley physiographic province. The set of pattern variables

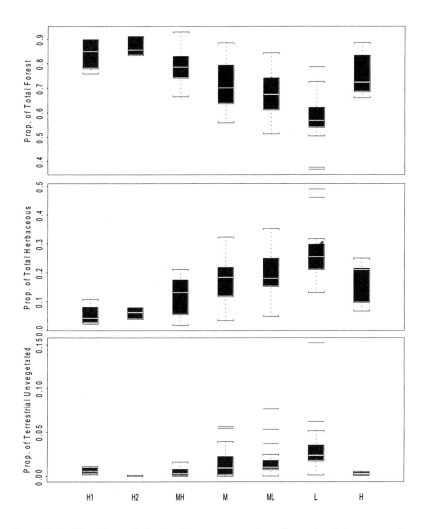

Figure 6.8. Box plots of the land cover proportions for each of the seven clusters identified in Figure 6.4. The general integrity of each cluster is labeled as follows: H1 and H2 = highest, MH = medium–high, M = medium, ML = medium–low, L = lowest and H = heterogeneous. The horizontal axis shows integrity from highest to lowest, except that the "heterogeneous" category (H) cannot be conveniently located along an integrity gradient.

in Figure 6.7 and the land cover summaries in Figure 6.8 constituted the initial set of potential predictors from which an optimal subset was chosen. Although the land cover proportions were correlated among themselves and to some degree with the spatial pattern variables, they were all included initially so that the stepwise regression procedure could pick and choose among them. Prior to applying the stepwise protocol, a log transform was applied to the proportion of "terrestrial unvegetated" land in order to mitigate the influence of an extreme outlier (refer to category "L" in the lower diagram of Figure 6.8.

The sum of high and highest integrity land yielded a poor response variable for linear modeling, as evidenced by unacceptable partial residual plots, other diagnostic plots and a very low R^2 of 0.37. Since pairwise plots of "high plus highest integrity land" with the potential predictors suggested that a log transform of the response may improve the fit, the response was transformed as log(response +1); however, an improvement was not seen and the R^2 remained essentially unchanged at 0.38.

When the proportion of low integrity BBA blocks was used as the response variable in the stepwise selection process, the resulting "optimal" set of regressor variables yielded diagnostic plots that were favorable with respect to a linear fit with approximately normally distributed residuals and no overly influential observations. The R^2 was only 0.57; however, besides an overall linear fit, the partial residual plots revealed some fairly strong linear relationships once the other predictors were factored into the model. These plots are seen in Figure 6.9, where we see that the landscape pattern variables that were retained, including both the A and B parameter estimates of the conditional entropy profiles, show fairly strong influence on the proportion of low integrity BBA blocks in a watershed. The computed linear coefficients are also reported in Figure 6.9 and summary statistics are reported in Table 6.3.

Table 6.3.   Coefficients and corresponding statistics for the optimal set of regressor variables for predicting the proportion of low integrity atlas blocks per watershed. Mean squared error (75 d.f.) = 0.97 and multiple $R^2 = 0.57$.

Regressor*	Value	t-value	p-value
Intercept	6.0968	2.9500	0.0042
App. Plateaus	−0.0831	−2.0196	0.0470
DLFD	−2.6258	−2.3951	0.0191
CONTAG	−0.0178	−4.9177	0.0000
A	−0.2812	−3.0365	0.0033
B	−1.1433	−2.9249	0.0046
TOT.HERB	0.6769	3.8952	0.0002

*Labels explained in Table 2.1 and Figure 4.1
App. Plateaus = identifies membership in Appalachian Plateaus

Finally, the raw BCI score, averaged for all blocks within a watershed, was compared to landscape variables by the stepwise regression proce-

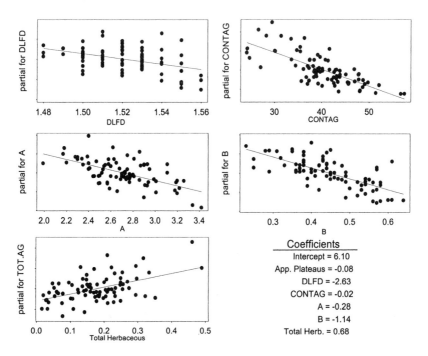

Figure 6.9. Partial residual plots from regressing the "proportion of low integrity BBA blocks" within each MAHA Pennsylvania watershed on the optimal set of landscape variables as listed in the lower right.

dure, resulting in the model defined in Table 6.4. The corresponding partial residual plots for the quantitative predictors, seen in Figure 6.10, reveal some strong linear relationships. The R^2 for this model is 0.69, which is somewhat stronger than for the model reported in Figure 6.9, especially since it contains 2 fewer predictor variables.

3.2   Clustering

Clustering of the Pennsylvania MAHA watersheds based on the spatial assessment of ecological integrity using the songbird community data (BCI proportions), as presented in Figures 6.4 and 6.5, was compared to clusters obtained from using all of the landscape measurements, as presented in chapter 4.

Comparisons for each of the physiographic provinces that overlap both Pennsylvania and the MAHA region are tabulated in Table 6.5. The cluster categories that were based solely on ecological integrity are the same for both physiographic provinces because this clustering is based on all watersheds across the two provinces; however, the cluster categories

Table 6.4. Coefficients and corresponding statistics from regressing the raw BCI score, averaged for all blocks within a watershed, against quantitative landscape variables and an indicator variable for specifying membership in one of two physiographic groups. Mean squared error (77 d.f.) = 13.7 and multiple R^2 = 0.69.

Regressor*	Value	t-value	p-value
Intercept	37.6907	24.6755	0.0000
App. Plateaus	1.1347	2.8318	0.0059
PSCV	−0.0002	−2.0926	0.0397
CONTAG	0.2219	5.8799	0.0000
TOT.HERB	−11.6307	−5.2991	0.0000

*Labels explained in Table 2.1
App. Plateaus = identifies membership in Appalachian Plateaus

that were based solely on landscape variables are somewhat different for each province because clustering was applied separately within each province due to very different fragmentation patterns.

These results indicate some fairly strong dependence between the two classification schemes. A formal chi–square or likelihood ratio test is not possible for either physiographic province because of the zero entries in their respective contingency tables; however, these zero entries are the result of a dependent structure.

For the Appalachian Plateaus, the landscape measurements did an excellent job of identifying "medium–low" watersheds, as evidenced by complete correspondence with clusters that were qualified as having "medium" to "medium–low" ecological integrity, and with the two watersheds that were qualified as having the "lowest" integrity. The landscape measurements also did a good job of identifying the "medium–high" and "highest" watersheds, as evidenced by distributions across the ecological integrity clusters that were biased towards the "medium–high" and "highest" clusters, respectively. However, one "medium–low" and one "lowest" integrity watershed was each misclassified as "highest" with the landscape variables.

For the Ridge and Valley, the landscape measurements identified the single watershed that was qualified as having the "highest" ecological integrity (Clarke Creek). A "medium" integrity watershed was also grouped with the highest watershed, but no gross misclassifications into "medium–low" or "lowest" general integrity occurred. The "medium–low" and "lowest" integrity watersheds of the Great Valley section were clearly identified by the landscape variables alone. Finally, out of the 21 watersheds identified as "medium" by the landscape variables, all but 4

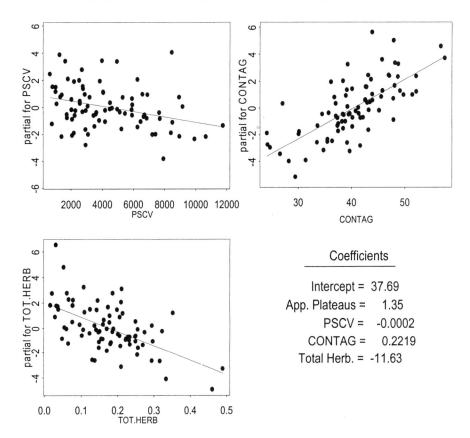

Figure 6.10. Partial residual plots from regressing the songbird-based bird community index, averaged across blocks within each MAHA watershed, on the landscape variables listed in the lower right.

were qualified as being in one of the medium–integrity categories, with 3 out of these 4 being labeled as lowest–integrity.

3.3 Conditional Entropy Profiles

Conditional entropy profiles are reported in Figure 6.11 for those watersheds coded as highest, medium and lowest in Figure 6.5. The medium-high and medium-low watersheds were not included in order to maintain better graphical clarity.

One observation is that the "highest" integrity watersheds lie near the bottom of the profiles. This is because the "highest" watersheds are mostly forested and therefore characterized by large coherent forest patches. This highly contagious pattern results in a very low conditional

Table 6.5. Comparison of watershed clusters obtained from BCI proportions to an independent clustering obtained from landscape measurements.

	Appalachian Plateaus						
LANDSCAPE	**BCI proportions**						
MEASUREMENTS	H†	MH	M	ML	L	Het	total
high	4	3	3	1	1	1	13
MH	2	5	6	3	0	1	17
ML	0	0	8	7	0	0	15
ML(Pitt.)*	0	0	2	2	2	0	6
total	6	8	19	13	3	2	51
	Ridge and Valley						
LANDSCAPE	**BCI proportions**						
MEASUREMENTS	H	MH	M	ML	L	Het	total
high	1	0	1	0	0	0	2
M (lower App. Mtns.)	0	2	2	7	0	2	13
M (upper App. Mtns.)	0	0	2	2	3	1	8
ML(lower Gr. Val.)	0	0	0	0	2	0	2
ML(upper Gr. Val.)	0	0	0	1	3	0	4
very low	0	0	0	0	2	0	2
total	1	2	5	10	10	3	31

† Symbols for qualitative cluster labels are:
H = highest, MH = medium–high, M = medium,
ML = medium–low, L = lowest, Het = heterogeneous,
* Symbols for geographic location, where appropriate are:
Pitt. = Pittsburgh and near vicinity,
"lower App. Mtns." and "upper App. Mtns." = the lower
(southwest) and upper (northeast) Appalachian
Mountains, respectively, and
"lower Gr. Val." and "upper Gr. Val." = the lower
(southwest) and upper (northeast) Great Valley, respectively

entropy near the floor resolution. Furthermore, these mostly forested watersheds contain some cover, such as terrestrial unvegetated land, that occurs rarely and in fine-grained patches. Such land cover types are "washed out" rapidly from the resampling filter used to degrade map resolution, therefore reducing the overall maximum conditional entropy that can be obtained as the resolution is increasingly degraded.

The profiles migrate upwards as the underlying landscape becomes increasingly fragmented and land cover becomes more evenly distributed both marginally and spatially. The increasing fragmentation generally corresponds with decreasing ecological integrity; however, the "lowest" integrity watersheds do not yield the top profiles. This is because wa-

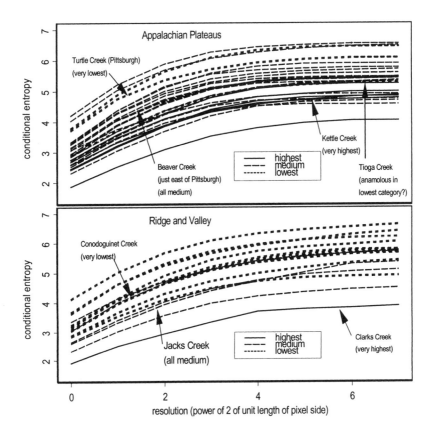

Figure 6.11. Conditional entropy profiles of watersheds in the MAHA section of Pennsylvania that were labeled as having "highest", "medium" or "lowest" general ecological integrity, as assessed from the songbird community.

tersheds of the very lowest integrity are those that have returned to a more contagious condition, but now it is due to large non-forest patches.

Within Figure 6.11, the "very lowest" watershed was chosen for each province as the one with the highest proportion of low-integrity blocks, the "very highest" had the highest proportion of high- plus highest-integrity blocks, and the "all medium" watersheds were covered 100% by medium-integrity blocks. This shows that as we go from the very highest to all medium to the very lowest watersheds, the profiles rise for all resolutions; however, when all watersheds are viewed, it appears that as the forest is increasingly fragmented, the maximum conditional entropy profile reflects a sort of critical state, beyond which the landscape matrix is non-forest and ecological integrity gets increasingly lower.

4. Ecological Integrity based on All Species in the Breeding Bird Atlas

An alternative approach to that taken in Section 2 for assessing ecological integrity is to use all of the atlas blocks so that an assessment could be made for every watershed. Since the MAHA songbird calibrations are not applicable outside of MAHA or to other species and guilds, we developed a new basis for categorizing watersheds applicable across the entire state of Pennsylvania. A set of 34 guilds incorporating species from the Breeding Bird Atlas was used (Bishop, 2000). Using these guilds and those species listed in the atlas as possible, probable or confirmed, two variations on a common protocol were investigated.

Reference watersheds were chosen to represent the extremes (best and worst) of ecosystem condition, based on the water quality study in Chapter 5 and the MAHA songbird evaluation in Section 2 of this chapter. Three watersheds from Pennsylvania's northern tier were chosen to represent the best condition since they had the highest PPI rankings (Chapter 5) and also the highest ecological integrity (Section 2). Four watersheds from the Piedmont Plateau and Great Valley section were chosen for the worst condition. These had the lowest PPI rankings; only one of these four fell within the MAHA region (the one in the Great Valley) but it had the lowest ecological integrity rating.

For each of the two reference groups of watersheds, a 34-guild mean vector was obtained by averaging the species proportions in each guild over all atlas blocks in the group of watersheds. This included 274 blocks for the three best watersheds and 166 blocks for the four worst watersheds. Given the reference mean vectors for the best ($\mathbf{P_b}$) and worst ($\mathbf{P_w}$) watershed groups, a vector of weights was established by subtraction: $\mathbf{D} = \mathbf{P_b} - \mathbf{P_w}$. Therefore, each guild, labeled as $j = 1, \cdots, G$, is represented by a weight $\mathbf{D}_j \in [-1, 1]$, so that larger (positive) weights indicate higher integrity and smaller (negative) weights indicate lower integrity.

Now, a bird community index can be calculated for each of $i = 1, \cdots, N$ atlas blocks across the state by the following sum:

$$\text{BCI}_i = \sum_{j=1}^{G} P_{ij} D_j, \tag{6.1}$$

where, for block i, P_{ij} is the proportion of species in the jth guild.

Two different sets of P_{ij} values were studied: first, P_{ij} was computed as the proportion of species in guild j out of all species in block i; secondly, the P_{ij} was computed as the proportion of species in guild j out of all potential species in guild j. The first approach, coined "species richness-based" follows the method used for the MAHA songbirds in section 2. A possible limitation of this approach is that the vector of species proportions can be identical for two different blocks, even though the two species richness totals may be very different. For this reason, the second approach, coined "guild potential-based" was considered.

Results for the two ways of calculating P_{ij} are presented in Figures 6.12 and 6.13, where the block values of BCI are averaged to obtain a summary BCI value for each watershed. Since there are some broad-scale statewide patterns, the state was stratified by a combination of physiographic provinces and sections to delineate more homogeneous subregions of the state using natural boundaries. These strata are incorporated into regression models that appear later in this Chapter.

Unlike with the MAHA songbirds, there are no established standards for assigning BCI scores to categories of ecological integrity. One may consider clustering the blocks in guild-space; however, with close to 5000 blocks in the state, analysis and interpretation would be very difficult. Accordingly, ecological integrity was represented by the BCI scores themselves.

4.1   Relation to Landscape Attributes

Since there is no basis for translating the BCI scores into ecological integrity categories, as was done with the MAHA songbirds in Section 3.1, then a spatial distribution of ecological integrity is not obtainable. Losing this spatial information is a major sacrifice; however, the average BCI scores do lend themselves to regression modeling.

The stepwise selection procedure, described in Section 3.1, was applied to the same set of potential quantitative predictor variables in order to assess the strength of relationship between the average BCI value and the landscape variables for the watersheds. Dummy variables for indicating physiographic membership of each watershed were forced to be retained by the stepwise protocol. Results for each of the two methods of calculating P_{ij} yielded favorable diagnostics. Furthermore, partial residual plots for the quantitative predictors, presented in Figures 6.14 and 6.15, indicate strong linear relationships. Summary statistics for each predictor are reported in Tables 6.6 and 6.7.

The "species richness-based" method of computing BCI yielded the best diagnostics and an R^2 of 0.78 (8 parameters). The "guild potential-based" method yielded an R^2 of 0.66 (7 parameters). These results indicate that a fairly strong relationship is obtained with the "species richness-based" method of computing BCI. This is supported by results in Table 6.4 where an R^2 of 0.69 is noted when the raw BCI values from the MAHA songbirds were regressed against only 5 parameters.

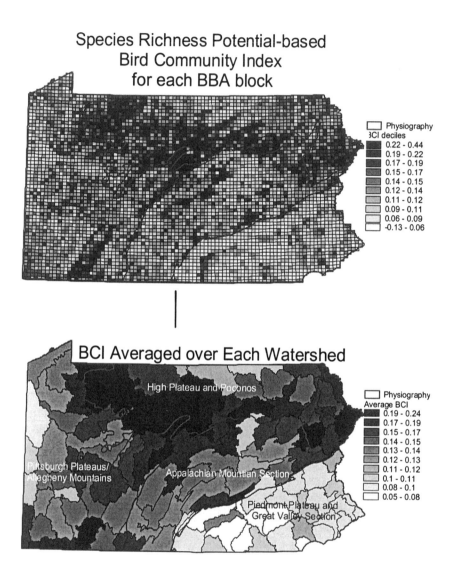

Figure 6.12. "Species richness-based bird community index" of ecological integrity, presented within each breeding bird atlas block (top) and for blocks averaged within each watershed. A combination of physiographic provinces and sections have been overlaid to stratify the state into more homogeneous areas of ecological integrity.

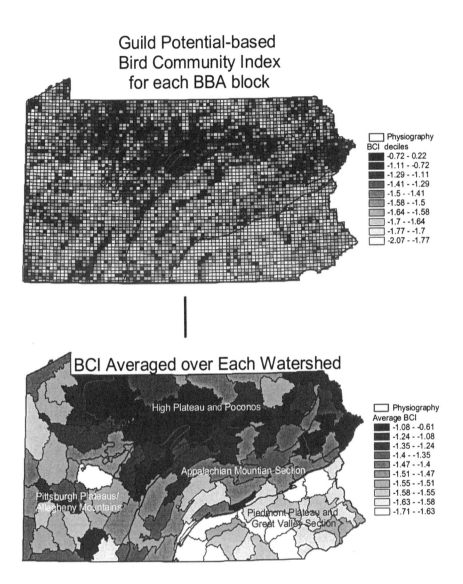

Figure 6.13. "Guild potential-based bird community index" of ecological integrity, presented within each breeding bird atlas block (top) and for blocks averaged within each watershed. A combination of physiographic provinces and sections have been overlaid to stratify the state into more homogeneous areas of ecological integrity.

Table 6.6. Coefficients and corresponding statistics from regressing the species richness-based BCI score, averaged for all blocks within a watershed, against quantitative landscape variables and indicator variables for specifying membership in a physiographic group. Applied to all watersheds in Pennsylvania. Mean squared error (94 d.f.) = 0.19 and multiple R^2 = 0.78.

Regressor*	Value	t-value	p-value
Intercept	−0.3204	−1.4741	0.1438
App.Mountain	−0.0276	−3.7969	0.0003
HighPlat.Pocono	−0.0038	−0.5209	0.6037
Pied.GrValley	−0.0368	−5.1539	0.0000
DLFD	0.2286	1.6553	0.1012
CONTAG	0.0015	3.5774	0.0006
B	−0.0701	−2.2483	0.0269
TOT.FOREST	0.1414	5.9101	0.0000

*Labels explained in Table 2.1 and Figure 4.1
App.Mountain = Appalachian Mountains Section
HighPlat.Pocono = High Plateaus Sections and Pocono Area
Pied.GrValley = Piedmont Plateau and Great Valley Section

Table 6.7. Coefficients and corresponding statistics from regressing the guild potential-based BCI score, averaged for all blocks within a watershed, against quantitative landscape variables and indicator variables for specifying membership in a physiographic group. Applied to all watersheds in Pennsylvania. Mean squared error (95 d.f.) = 1.33 and multiple R^2 = 0.66.

Regressor*	Value	t-value	p-value
Intercept	−1.1653	−7.0529	0.0000
App.Mountain	−0.1277	−2.6240	0.0101
HighPlat.Pocono	0.1063	2.2922	0.0241
Pied.GrValley	−0.1071	−2.4600	0.0157
CONTAG	0.0054	2.1643	0.0329
B	−0.5351	−2.6726	0.0089
TOT.HERB	−0.8786	−5.1467	0.0000

*Labels explained in Table 2.1, Figure 4.1 and Table 6.6

5. Comparison of the Different Methods for Computing a Bird Community Index

Consistency among the three different approaches presented in this Chapter for computing a bird community index of ecological integrity is evaluated in Table 6.8. For each of the watershed clusters that were determined from landscape measurements alone, as discussed in Chapter

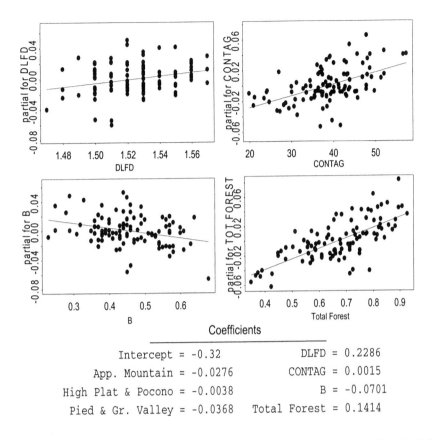

Figure 6.14. Partial residual plots from regressing the "species richness-based bird community index" (BCI) on the landscape variables listed, for all watersheds in Pennsylvania.

4, the average BCI was obtained by taking an area–weighted average BCI of all watersheds in each respective cluster. Table 6.8 does indeed show that for the majority of Pennsylvania (all of the Appalachian Plateaus and the Ridge and Valley), the average BCI consistently increased for all three methods of computing the BCI as the watershed clusters went from lowest to highest amount of forest cover.

In the Piedmont, the two methods that could be applied did not reveal consistent results and they also did not coincide with apparent decreasing forest cover. This physiographic region, however, presents several limitations for comparing average BCI values among landscape-based watershed clusters. First, there are no watersheds that are mostly forested to provide a background. The one watershed that is labeled

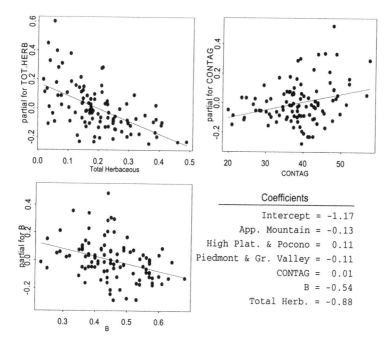

Figure 6.15. Partial residual plots from regressing the "guild potential-based bird community index" (BCI) on the landscape variables listed, for all watersheds in Pennsylvania.

as "medium–to–high" with respect to forest cover is actually not much different than those in the "medium" cluster. Secondly, the Piedmont is a small area of Pennsylvania and therefore the sample of watersheds is much smaller than for the other two provinces.

6.   Summary

Landscape variables provided fairly strong predictors for each of three different approaches to obtaining a watershed–wide average bird community index (BCI) of ecological integrity. After factoring out physiographic regimes, an objective stepwise model selection procedure retained landscape variables that explained 66% to 78% of the variation in BCI. As expected, for each method of obtaining the BCI response, either total forest or total herbaceous land served as a very strong predictor; however, several measurements of spatial pattern were also retained. Of these pattern measurements, the conditional entropy term that relates the rate of information loss (parameter B) was retained twice and contagion was retained all three times. Contagion was highly significant in

Table 6.8. Area-weighted bird community indices (BCIs) for three different protocols within each cluster defined by landscape measurements.

province	clusters based on landscape measurements	Area–Weighted BCI Based On		
		species richness	guild potential	MAHA songbirds
AP	high (13)*	0.189	−1.019	47.606
	MH† (17)	0.163	−1.243	46.045
	ML1 (6)	0.132	−1.437	44.310
	ML2(19)	0.115	−1.443	42.302
RV	high (2)	0.167	−1.349	46.223
	M1(13)	0.136	−1.452	43.767
	M2 (8)	0.134	−1.391	42.924
	ML1 (2)	0.097	−1.606	40.258
	ML2 (4)	0.091	−1.626	40.466
	very low (2)	0.076	−1.678	39.215
Pied.	MH (1)	0.046	−1.600	N/A
	M (4)	0.084	−1.570	N/A
	ML (4)	0.066	−1.632	N/A
	very low (7)	0.086	−1.630	N/A

* () = number of watersheds in cluster
† Symbols for qualitative cluster labels are:
MH = medium–to–high, M = medium, ML = medium–to–low.
Within the Appalachian Plateaus (AP),
ML2 = Pittsburgh and near vicinity, while
ML1 = medium–to–low elsewhere.
Within the Ridge and Valley (RV),
M1 and M2 refer to medium watersheds in the lower
(southwest) and upper (northeast) Appalachian
Mountains, respectively, and
ML1 and ML2 refer to medium–to–low watersheds in the
lower (southwest) and upper (northeast) Great Valley, respectively.

all cases and revealed a fairly strong linear relationship to BCI. Since contagion is very highly linearly correlated with the conditional entropy term C, this indicates that using only land cover proportions and terms from the multi–resolution conditional entropy profiles (A, B and C in Figure 4.1) can provide a capability to predict general landscape-scale ecological integrity.

For the songbird-based approach to obtaining BCI values within the MAHA region, the watersheds could be clustered based on their respective proportions of breeding bird atlas blocks in "low", "medium", "high" and "highest" integrity. When these watershed clusters were compared to distinct clusters obtained using only landscape variables, a high degree of correspondence was observed, thus further supporting the claim that landscape measurements alone can provide a fairly strong assessment of landscape-level ecosystem condition.

Developing a landscape-level index of biotic integrity, whether based on breeding birds or other communities, is an ongoing process that will be refined and improved with further learning. For example, the two approaches to computing a watershed–wide BCI that were presented in Section 4 may be improved by reducing the full set of guilds to remove redundancy prior to computing a BCI. Also, one can experiment with different choices of reference watersheds.

With a new breeding bird atlas underway, there will be interesting opportunities for temporal comparisons.

Chapter 7

SUMMARY AND FUTURE DIRECTIONS

Research presented in this monograph helps advance the science of landscape ecology in several ways. Conditional entropy profiles were developed and illustrated as a multi-resolution method of quantifying pattern in land cover maps that captures multi-scale characteristics of pattern. Finally, the ability of landscape measurements to predict different aspects of ecosystem condition was evaluated using independent characterizations of watershed-wide surface water pollution and ecological integrity. Some fundamental questions and findings are now further elaborated.

Are conditional entropy profiles sensitive to changing spatial patterns and can they distinguish different landscape types?

For Pennsylvania, located in a temperate zone with a potential natural vegetation of predominantly hardwood (Mikan, Orwig and Abrams, 1994; Nowacki and Abrams, 1992) or mixed conifer/hardwood (Whitney, 1990) forest, the conditional entropy profiles do provide a strong graphical tool with a sound quantitative basis for characterizing and monitoring landscapes. However, this assessment of spatial pattern needs to be accompanied by the non-spatial marginal land cover proportions for best discriminatory power.

When plotted on a common scale, profiles obtained from different landscape patterns reveal distinctly different responses. For profiles that reflect similar patterns, meaning that they cross at some resolution(s) or lie very close, the marginal land cover proportions help clarify whether the underlying landscapes are truly similar or just have similar spatial patterns. This would be necessary because conditional entropy (as with contagion and the interspersion and juxtaposition index) is label independent. For example, if a landscape is dominated by large coherent patches, the conditional entropy could be a certain value regardless of whether those patches are forest or grassland. A great value of conditional entropy profiles is that they summarize complex spatial patterns and incorporate multiple measurement resolutions in a way that can be readily visualized in a two-dimensional graph.

What advantage is there to conditional entropy over other established measurements?

It is reasonable to argue that similar multi-resolution profiles can be obtained for other more conventional landscape measurements; however, there are distinct advantages to the conditional entropy profile approach.

When re-scaling a raster map through a random filter, expected 4-tuple frequencies can be directly computed for multiple resolutions, given only the floor resolution data (see Chapter 2); therefore conditional entropy profiles can be obtained for these expected frequencies, which is much more meaningful than obtaining one realization or even a sample of realizations from particular runs of a re-scaling filter. Furthermore, the conditional entropy profiles are directly related to many other measurements, as seen by the rather precise linear correlation of the maximum conditional entropy with contagion, which is in turn highly correlated with most other single-resolution measurements (see Figures 4.2 and 4.3 in Chapter 4).

Can landscape-level ecosystem condition be reliably predicted primarily from landscape measurements on land cover data derived from remotely-sensed imagery?

For these Pennsylvania watersheds, landscape variables did indeed explain much of the observed variation of surface water pollution loading, whether represented by field-measured nutrient loading or a modeled pollution potential index. After factoring out physiographic associations to minimize spatial autocorrelation, an objective stepwise model selection protocol retained several landscape predictors for both types of response variables. The resulting linear models each explained 76% of the variation in the respective water pollution variables. (It is purely coincidence that the coefficient of multiple determination, or R^2, rounds to 0.76 for each model.) The proportion of annual herbaceous land cover was the strongest predictor of total nitrogen loading and the proportion of total herbaceous land (annual plus perennial) was the strongest predictor of the pollution potential index. These results are certainly not surprising; however, several variables that measure spatial *pattern* were also retained because they improved the predictability of each of the two responses. Furthermore, of the pattern variables that were retained, some aspect of the multi-resolution conditional entropy profiles were always included. Cooper, Helliwell and Coull (2004) have since presented ways to improve catchment selection and modelling for using landscape and other variables to predict stream characteristics such as acid neutralizing capacity.

For predicting ecological integrity, landscape variables provided fairly strong predictors for each of three different approaches to obtaining a watershed-wide average bird community index (BCI) of ecological integrity. After factoring out physiographic regions, an objective stepwise model selection procedure retained landscape variables that explained 66% to 78% of the variation in BCI. As expected, either total forest or total herbaceous land served as a very strong predictor of BCI; however, several measurements of spatial pattern were also retained. Of these pattern measurements, the conditional entropy term that relates the rate of information loss (parameter B) was retained twice and contagion was retained all three times. Contagion was highly significant in all cases and revealed a fairly strong linear relationship to BCI. Since contagion is very highly linearly correlated with maximum conditional entropy (term C), this indicates that land cover (either total forest or total herbaceous) along with the multi-resolution conditional entropy terms (A, B and C in Figure 4.1) can provide a strong capability to predict general landscape-scale ecological integrity.

For the songbird-based approach to obtaining BCI values within the MAHA region, the watersheds could be clustered based on their respective proportions of breeding bird atlas blocks in "low", "medium", "high" and "highest" integrity. When these watershed clusters were compared to distinct clusters obtained using only landscape variables, a high degree of correspondence was observed, thus further supporting the claim that landscape measurements can characterize landscape-level ecosystem condition.

Future Directions

Identify Landscapes that are at Critical Transitions:

It would be of utmost importance if one could discover the values for some set of landscape measurements that identified a landscape-level ecosystem to be at a critical transition point, whereby the ecosystem is moving from a healthy to unhealthy phase.

This would be akin to measuring the critical proportion of forest cover in a forest/non-forest landscape that equals the threshold at which the forest percolates through the landscape (Gardner, Milne, Turner and O'Neill, 1987). For organisms that require full connectivity of the forest across the landscape, breaking the connectivity implies their demise. While there is a well known percolation threshold for random maps, actual landscapes present an altogether different situation. For example, real landscapes have exhibited forest percolation at proportions much less than the theoretical percolation threshold for a random map, and this has been attributed to the hierarchical structure of landscapes (Lavoral, Gardner and O'Neill, 1993). A further complication is that a landscape is defined by many land cover types, not just forest versus non-forest. So, it does not take long to see that identifying some aspect of a landscape that identifies it to be at a critical transition may be too evasive due to intractable complexities; however, all research along the lines of this monograph should bear such an objective in mind.

Of course, in order to evaluate critical landscape transitions, one needs to define what aspect of the ecosystem is of concern. Is it overall species richness or some measurement of species diversity? Perhaps species richness is increasing although keystone species populations are declining. The bird community indices established in Chapter 6 aim to establish a more meaningful index of biotic integrity that generally reflects the degree of disturbance of native ecosystem attributes.

Once a desired ecosystem response is identified, one then needs to clarify what constitutes healthy versus unhealthy and consequently when an ecosystem is in a critical transition. If a background standard can be established that represents a pristine ecosystem, or at least the healthiest condition in a particular region, then some type of distance could be measured from such a background standard. As other landscape-level ecosystems become more ecologically "distant" from background, this would reflect decreasing health.

As an initial exploratory exercise, the Pennsylvania watersheds studied in this monograph were represented by the Euclidean distance of their respective guild mean vectors (for the full 34 guilds) from the background vector of guild means discussed in Section 4 of Chapter 6. The resulting trend of increasing distance from background did not reveal distinct transition points; however, there were two smooth regions of inflection. When the conditional entropy profiles for the watersheds in these inflection regions were evaluated, there were no consistent properties of the profiles. Further analysis of other landscape pattern properties could now be done. For example, the value of different landscape measurements could be plotted as a function of the order of ecological distance to see if there are any common trends with the trend in ecological distance. Furthermore, other measures of distance, such as Mahalanobis Distance, can be evaluated.

Evaluate Conditional Entropy Profiles at other Landscape Extents

While conditional entropy profiles proved useful for watersheds of the given scale within Pennsylvania, they remain to be evaluated on landscapes of much larger or smaller spatial extents, and also from other climatic and physiographic regimes.

Determine Significance of Spatial Pattern Versus Simple Land Cover Proportions for Different Landscape Extents

It has been observed that variations in surface water quality for small watersheds, especially those for headwater streams, can be explained by both the marginal land cover proportions *and* the spatial pattern of land cover, whereas for larger watersheds, marginal land cover proportions alone can explain water quality variability. The watersheds used in this monograph revealed that both proportions and spatial pattern were significant factors affecting both water quality and ecological integrity. It is now of interest to see what landscape factors are significant for both smaller and larger watersheds. The questions that arise include "Is there a particular scaling level for the size of a watershed beyond which pattern is not a useful predictor of water quality?" and "Is this scaling level consistent around the earth, or within particular regions?".

Further Opportunities with the Breeding bird Atlas

The analysis in Section 4 of Chapter 6 actually just establishes a first phase of analyzing the full breeding bird atlas for assessing ecological condition for the watersheds or other defined regions. For any particular region where it is desired to perform an ecological assessment, a more detailed approach can be taken. First, all the guilds can be analyzed for inter-correlations and a smaller set of guilds can be chosen that does not contain redundant information. The selected set of guilds can then be used to cluster blocks within the region of interest and an approach similar to that by O'Connell, Jackson and Brooks (1998a) could be taken to eventually categorize all the blocks in the region to be in some category of ecological integrity. If such a protocol was to be applied to all the watersheds studied in this monograph, it would be a formidable task; however, one could also apply the protocol to the whole state, or perhaps in each of the three major physiographic provinces. What then becomes formidable is to have to cluster so many blocks, with nearly 5000 covering the state. This problem may be addressed by dealing with smaller samples of watersheds so that one is only clustering the blocks for these particular watersheds.

A new atlas effort is underway for Pennsylvania, which should provide interesting opportunities for comparison, especially if combined with atlases of neighboring states.

Additional Work at the Press Time

Consistent with the theme of this monograph, the material in the following references will be a healthy reinforcement to the contents of the monograph. See Burnicki, A.C. (2001); Grossi, L., Patil, G.P. and Taillie, C. (2004); Modarres, R. and Patil, G.P. (2006); Myers, W.L., Kong, N. and Patil, G.P. (2005); Myers, W.L., Kurihara, K., Patil, G.P. and Vraney, R.(2006); Myers, W.L., McKenney-Easterling, M., Hychka, K., Griscom, B., Bishop, J., Bayard, A., Rocco, G., Brooks, R., Constantz, G., Patil, G.P. and Taillie, C.(2006); Myers, W. L. and Patil, G. P. (2006); Myers, W.L., Patil, G.P. and Cai, Y. (2006); Patil, G.P., (2002, 2003, 2004, 2005, 2006ab); Patil, G.P., Acharya, R., Glasmeier, A., Myers, W.L., Phoha, S. and Rathbun, S. (2007); Patil, G.P., Acharya, R., Modarres, R., Myers, W.L. and Rathbun, S.L. (2006); Patil, G.P., Balbus, J.A., Biging, G., JaJa, J., Myers, W.L. and Taillie, C. (2004); Patil, G.P., Bishop, J.A., Myers, W.L., Taillie, C., Vraney, R. and Wardrop, D. (2004); Patil, G.P., Brooks, R.P., Myers, W.L. and Taillie, C. (2003); Patil, G.P., Duczmal, L., Haran, M. and Patankar, P. (2006); Patil, G.P., Johnson, G.D., Taillie, C. and Myers, C. (2000); Patil, G.P., Modarres, R., Myers, W.L. and Patankar, P. (2006); Patil, G.P, Myers, W.L, Luo, Z., Johnson, G.D. and Taillie, C. (2000); Patil, G.P. and Taillie, C. (1999); Patil, G.P. and Taillie, C. (2002); Patil, G.P. and Taillie, C., (2004ab); Rapport, D.J., Lasley, W.L., Rolston, D.E., Neilsen, N.O., Qualset, C.O. and Damania, A.B. (2003); Rathbun, S. and Black, B. (2006); Rathbun, S. and Fei, S. (2006); and Rodríguez, S. (2001).

Chapter 8

REFERENCES

Akaike, H. 1974. A new look at statistical model identification. *IEEE Transactions on Automatic Control* AU-19, pp. 716–722.

Askins, R.A. 1995. Hostile landscapes and the decline of migratory songbirds. *Science*, 267:1956–1957.

Aspinall, R. and Pearson, D. 2000. Integrated geographical assessment of environmental condition in water catchments: Linking landscape ecology, environmental modelling and GIS. *J. Environmental Management*, 59:299–319.

Basharin, G.P. 1959. On a statistical estimate for the entropy of a sequence of independent random variables. Theory of Probability and its Applications, 4:333–336.

Beier, P. and Noss, R.F. 1998. Do habitat corridors provide connectivity? *Conservation Biology*, 12(6):1241-1252.

Benson, B.J and MacKenzie, M.D. 1995. Effects of sensor spatial resolution on landscape structure parameters. *Landscape Ecology*, 10(2):113–120.

Bishop, J.A. 2000. Associations between avian functional guild response and regional landscape properties for conservation planning. M.S. Thesis. The Pennsylvania State University.

Bradford, D.F., Franson, S.E., Miller, G.R., Neale, A.C., Cantebury, G.E. and Heggem, D.T. 1998. Bird species assemblages as indicators of biological integrity in Great Basin rangeland. *Environmental Monitoring and Assessment*, 49:1–22.

Brauning, D.W. 1992. *Atlas of Breeding Birds in Pennsylvania.* University of Pittsburgh Press, Pittsburgh, Pennsylvania.

Brauning, D. W. and F. B. Gill. 1983–1989. *Pennsylvania Breeding Bird Atlas data.* The Academy of Natural Sciences of Philadelphia, Pennsylvania Game Commission, and Wild Resource Conservation Fund, Harrisburg, PA.

Brooks, T. and Kennedy, E. 2004. Biodiversity barometers. *Nature*, 431:1046–1047.

Brooks, R.P., O'Connell, T.J., Wardrop, D.H. and Jackson, L.E. 1998. Towards a Regional Index of Biological Integrity: The Example of Forested Riparian Ecosystems. *Environmental Monitoring and Assessment*, 51:131–143.

Brittingham, M.C. and Temple, S.A. 1983. Have cowbirds caused forest songbirds to decline? *Bio–Science*, 33:31–35.

Burnicki, A. C. 2001. Comparison of absolute and conditional entropy profiles for assessing landscape fragmentation. M.Sc. Thesis, The Pennsylvania State University, University Park, PA.

Carlsen, T.A., Coty, J.D. and Kercher, J.R. 2004. The spatial extent of contaminants and the landscape scale: an analysis of the wildlife, conservation biology and population literature. *Environmental Toxicilogy and Chemistry*, 23(3):798–811.

Collins, J.B. and Woodcock, C.E. 1996. Explicit consideration of multiple landscape scales while selecting spatial resolutions. In: Mowrer, H.T., Czaplewski, R.L. and Hamre, R.H. (eds.) *Spatial Accuracy Assessment in Natural Resources and Environmental Sciences*, Second International Symposium. USDA Forest Service, Rocky Mountain Forest and Range Experimental Station, Fort Collins CO. General Technical Report RM-GTR-277.

Colwell, R.K. 1974. Predictability, constancy, and contingency of periodic phenomena. *Ecology*, 55:1148–1153.

Cooper, D.M., Helliwell, R.C. and Coull, M.C. 2004. Predicting acid neutralizing capacity from landscape classificaion: application to Galloway, south-west Scotland. *Hydrological Processes*, 18:455-471.

Costanza, R. and Maxwell, T. 1994. Resolution and predictability: An approach to the scaling problem. *Landscape Ecology*, 9(1):47–57.

Daniel, C. and Wood, F.S. 1980. *Fitting Equations to Data*, 2nd Ed. Wiley, New York

DeSalle, R. and Amato, G. 2004. The expansion of conservation genetics. *Nature Reviews Genetics*, 5(9):702-712.

Dubes, R.C. and Jain, A.K. 1989. Random field models in image analysis. *J. Applied Statistics*, 16(2):131–165.

Ernoult, A., Bureau, F. and Poudevigne, I. 2003. Patterns of organisation in changing landscapes: implications for the management of biodiversity. *Landscape Ecology*, 18:239-251.

Forman, R.T.T. 1995. Some general principles of landscape and regional ecology. *Landscape Ecology*, 10(3):133–142.

Frankham, R. 2005. Ecosystem recovery enhanced by genotypic diversity. *Heredity*, 95:183.

Franklin, J.F. and Forman, R.T.T. 1987. Creating landscape patterns by forest cutting: Ecological consequences and principles. *Landscape Ecology*, 1(1):5–28.

Frohn, R.C. 1998. *Remote Sensing for Landscape Ecology; New Metric Indicators for Monitoring, Modeling and Assessment of Ecosystems.* Lewis, Boca Raton, 99 pp.

Gardner, R.H., Milne, B.T., Turner, M.G. and O'Neill, R.V. 1987. Neutral models for the analysis of broad-scale landscape pattern. *Landscape Ecology*, 1(1):19–28.

Gentry, A.H. 1996. Species expirations and current extinction rates: a review of the evidence. In: Szaro, R.C. and Johnston, D.W. (eds.), *Biodiversity in Managed Landscapes, Theory and Practice.* Oxford University Press, New York.

Graham, R.L., Hunsaker, C.T., O'Neill, R.V. and Jackson, B.L. 1991. Ecological risk assessment at the regional scale. *Ecological Applications*, 1(2):196–206.

Grossi, L., Patil, G.P., and Taillie, C. 2004. Statistical selection of perimeter-area models for patch mosaics in multiscale landscape analysis. *Environmental and Ecological Statistics*, 11(2):165-181.

Grumbine, R.E. 1994. What is ecosystem management? *Conservation Biology*, 8(1):27–38.

Gustafson, E.J, and Parker, G.R. 1992. Relationships between landcover proportion and indices of landscape spatial pattern. *Landscape Ecology*, 7:101–110.

Hamlett, J.M., Miller, D.A., Day, R.L., Peterson, G.W., Baumer, G.M. and Russo, J. 1992. Statewide GIS-based ranking of watersheds for agricultural pollution prevention. *J. Soil and Water Conservation*, 47(5):399–404.

Hanski, I. and Ovaskainen, O. 2000. The metapopulation capacity of a fragmented landscape. *Nature*, 404:755-758.

Hargis, C.D., Bissonette, J.A. and David, J.L. 1998. The behavior of landscape metrics commonly used in the study of habitat fragmentation. *Landscape Ecology*, 13:167–186.

Hastings, H.M. and Sugihara, G. 1993. *Fractals: A User's Guide for the Natural Sciences*. Oxford University Press, Oxford. 235 pp.

Hunsaker, C.T. and Levine, D.A. 1995. Hierarchical approaches to the study of water quality in rivers. *BioScience*, 45(3):193–203.

Johnson, D.H., Igl, L.D. and Dechant Shaffer, J.A. (Series Coordinators). 2004. Effects of management practices on grassland birds. Northern Prairie Wildlife Research Center, Jamestown, ND. Jamestown, ND: Northern Prairie Wildlife Research Center Online. http://www.npwrc.usgs.gov/resource/literatr/grasbird/grasbird.htm (Version 12AUG2004).

Johnson, G.D., Myers, W.L and Patil, G.P. 2001. Predictability of surface water pollution loading in Pennsylvania using watershed-based landscape measurements. *J. American Water Resources Association*, 37(4):821–835.

Johnson, G.D., Myers, W.L and Patil, G.P., 1999. Stochastic generating models for simulating hierarchically structured multi-cover landscapes. *Landscape Ecology*, 14:413–421.

Johnson, G.D., Myers, W.L, Patil, G.P. and Taillie, C. 1998. Quantitative characterization of hierarchically scaled landscape patterns. *1998 ASA Proceedings of the Section on Statistics and the Environment*, pp. 63–69.

Johnson, G.D., Myers, W.L, Patil, G.P. and Taillie, C. 1999. Multiresolution fragmentation profiles for assessing hierarchically structured landscape patterns. *Ecological Modeling*, 116:293–301.

Johnson, G.D., Myers, W.L, Patil, G.P. and Taillie, C. 2001a. Fragmentation profiles for real and simulated landscapes. *Environmental and Ecological Statistics*, 8(1):5–20.

Johnson, G.D., Myers, W.L, Patil, G.P. and Taillie, C. 2001b. Characterizing watershed-delineated landscapes in Pennsylvania using conditional entropy profiles. *Landscape Ecology*, 16(7):597–610.

Johnson, G.D., Myers, W.L., Patil, G.P., O'Connell, T.J., and Brooks, R.P. 2003. Predictability of bird community-based ecological integrity using landscape measurements. In: Rapport, D.J., Lasley, W.L., Rolston, D.E., Neilsen, N.O., Qualset, C.O., and Damania, A.B. (eds), *Managing for Healthy Ecosystems*, pp. 617–637. Lewis Publishers, Boca Raton, FL.

Johnson, G., Myers, W., Patil, G., and Walrath, D. 1998. Multiscale analysis of the spatial distribution of breeding bird species. In *Assessment of Biodiversity for Improved Forest Planning*, P. Bachmann, M. Kohl, and R. Paivinen, eds. Kluwer Academic Publishers, Dordrecht. pp. 135–150.

Johnson, G.D. and Patil, G.P. 1998. Quantitative multiresolution characterization of landscape patterns for assessing the status of ecosystem health in watershed management areas. *Ecosystem Health*, 4(3):177–187.

Johnson, G.D., Tempelman, A.K. and Patil, G.P. 1995. Fractal based methods in ecology: a review for analysis at multiple spatial scales. *Coenosis*, 10(2-3), 123-131.

Jones, K.B, Riitters, K.H., Wickham, J.D., Tankersley, R.D. Jr., O'Neill, R.V., Chaloud, D.J., Smith, E.R. and Neale, A.C. 1997. *An Ecological Assessment of the United States Mid-Atlantic Region*. USEPA, ORD, EPA/600/R-97/130.

Kaennel, M. 1998. Biodiversity: a diversity in definition. In *Assessment of Biodiversity for Improved Forest Planning*, P. Bachmann, M. Kohl, and R. Paivinen, eds. Kluwer Academic Publishers, Dordrecht.

Kaiser, J. 2001. Conservation biology: building a case for biological corridors. *Science*, 293(5538):2199.

Karr, J.R. 1991. Biological integrity: a long-neglected aspect of water resource management. *Ecological Applications*, 1:66–84.

Karr, J.R. 1993. Defining and assessing ecological integrity: beyond water quality. *Environmental Toxicology and Chemistry*, 12:1521–1531.

Keitt, T.H., Urban, D.L. and Milne, B.T. 1997. Detecting critical scales in fragmented landscapes. *Conservation Ecology*, 1(1):4. Available from the internet. URL: http://www.consecol.org/vol1/iss1/art4

Kotliar, N.B. and Wiens. 1990. Multiple scales of patchiness and patch structure: a hierarchical framework for the study of heterogeneity. *Oikos*, 59:253–260.

Krummel, J.R., Gardner, R.H., Sugihara, G., O'Neill, R.V. and Coleman, P.R. 1987. Landscape patterns in a disturbed environment. *Oikos*, 48:321–324.

Lavoral, S., Gardner, R.H. and O'Neill, R.V. 1993. Analysis of patterns in hierarchically structured landscapes. *Oikos*, 67:521–528.

Levin, S. 1992. The problem of pattern and scale in ecology. *Ecology*, 73(6):1943–1967.

Levin, S.A., Grenfell, B., Hastings, A. and Perelson, A.S. 1997. Mathematical and computational challenges in population biology and ecosystem science. *Science*, 275:334–343.

Li, H. and Reynolds, J.F. 1993. A new contagion index to quantify spatial patterns of landscapes. *Landscape Ecology*, 8(3):155–162.

Loehle, C. and Wein, G. 1994. Landscape habitat diversity: a multiscale information theory approach. *Ecological Modeling*, 73:311–329.

MacArthur, R.H. and Wilson, E.O. 1967. *The Theory of Island Biogeography. Monographs in Population Biology, 1*. Princeton University Press, Princeton, NJ.

Mallows, C.L. 1973. Some comments on Cp. *Technometrics*, 15:661–675.

MathSoft, Inc. 1997. *Splus 4 Guide to Statistics*. Data Analysis Products Division, MathSoft, Seattle. 877 pp.

McGarigal, K. and Marks, B. 1995. *FRAGSTATS: spatial pattern analysis program for quantifying landscape structure*. Gen. Tech. Rep. PNW-GTR-351. Portland, OR: U.S. Department of Agriculture, Forest Service, Pacific Northwest Research Station. 122 pp.

Maurer, B.A. 1994. *Geographical Population Analysis: Tools for the Analysis of Biodiversity.* Blackwell Scientific, London, 130 pp.

Mehaffey, M.H., Wade, T.G., Nash, M.S. and Edmonds, C.M. 2003. Analysis of land cover and water quality in the New York Catskill-Delaware basins. In: Rapport, D.J., Lasley, W.L., Rolston, D.E., Neilsen, N.O., Qualset, C.O., and Damania, A.B. (eds), *Managing for Healthy Ecosystems*, pp. 1327-1340. Lewis Publishers, Boca Raton, FL.

Mikan, C.J., Orwig, D.A. and Abrams, M.D. 1994. Age structure and successional dynamics of a presettlement-origin chestnut oak forest in the Pennsylvania Piedmont. *Bulletin of the Torrey Botanical Club*, 121(1):13–23.

Miller, E.W. 1995. *A Geography of Pennsylvania.* The Pennsylvania State University Press. University Park, PA. 406 pp.

Mladenoff, D.J. and DeZonia, B. 2001. APACK 2.17 Analysis Software. available at
http://landscape.forest.wisc.edu/Projects/APACK/apack.html

Modarres, R. and Patil, G.P. 2006. *Hotspot detection with bivariate data.* JSPI (S.N. Roy Centennial Volume). In Press.

Montgomery , D.C. and Peck, E.A. 1982. *Introduction to Linear Regression Analysis.* John Wiley and Sons, New York, 504 pp.

Mouquet, N. and Loreau, M. 2003. Community patterns in source-sink metacommunities. *The American Naturalist*, 162(5):544-557.

Myers, W. 1999. *Remote Sensing and Quantitative Geogrids in PHASES [Pixel Hyperclusters As Segmented Environmental Signals], Release 3.4.* Technical Report ER9710, Environmental Resources Research Institute, The Pennsylvania State University, University Park, PA. 57 pp. and 2 diskettes.

Myers, W.L., Kong, N., and Patil, G.P. 2005. Topological approaches to terrain in ecological landscape mapping. *Community Ecology*, 6(2):191-201.

Myers, W.L., Kurihara, K., Patil, G.P., and Vraney, R. 2006. Finding upper-level sets in cellular surface data using Echelons and SaTScan. *Environmental and Ecological Statistics*, 13(4). In Press.

Myers, W.L., McKenney-Easterling, M., Hychka, K., Griscom, B., Bishop, J., Bayard, A., Rocco, G., Brooks, R., Constantz, G., Patil, G.P., and Taillie, C. 2006. Contextual clustering for configuring collaborative conservation of watersheds in the Mid-Atlantic highlands.*Environmental and Ecological Statistics*, 13(4). In Press.

Myers, W. L. and Patil, G. P. 2006. *Pattern-Based Compression of Multi-Band Image Data for Landscape Analysis*. Springer, Boston, MA. In Press.

Myers, W.L., Patil, G.P., and Cai, Y. 2006. Exploring Patterns of Habitat Diversity Across Landscapes Using Partial Ordering. In: *Partial Order in Environmental Science and Chemistry*, Bruggemann, Rainer; Carlsen, Lars, eds. Springer, Boston, MA. In Press.

Neter, J., Wasserman, W. and Kutner, M.H. 1985. *Applied Linear Statistical Models, second ed.* Richard D. Irwin, Homewood, Ill., 1127 pp.

Nizeyimana,E, Evans, B.M., Anderson, M.C., Petersen, G.W., DeWalle, D.R., Sharpe, W.E., Hamlett, J.M. and Swistock, B.R. 1997. *Quantification of NPS Pollution Loads Within Pennsylvania Watersheds*. ER9708. Environmental Resources Research Institute, University Park PA.

Noss, R.F. 1983. A regional landscape approach to maintain diversity. *Bioscience*, 33(11):700–706.

Noss, R.F. 1996. Conservation of biodiversity at the landscape scale. In: Szaro, R.C. and Johnston, D.W. (eds.), *Biodiversity in Managed Landscapes, Theory and Practice*, pp. 574–589. Oxford University Press, New York.

Nowacki, G.J. and Abrams, M.D. 1992. Community, edaphic and historical analysis of mixed oak forests of the Ridge and Valley Province in central Pennsylvania. *Canadian J. Forest Research*, 22:790–800.

O'Connell, T.J., Jackson, L.E. and Brooks, R.P. 1998a. *The Bird Community Index: A Tool for Assessing Biotic Integrity in the Mid-Atlantic Highlands, Final Report*. Penn State Cooperative Wetlands Center, Report No. 98-4. Forest Resources Laboratory, Pennsylvania State University, University Park, PA 16802, USA. 57 pp.

O'Connell, T.J., Jackson, L.E. and Brooks, R.P. 1998b. A bird community index of biotic integrity for the Mid-Atlantic Highlands. *Environmental Monitoring and Assessment*, 51:145–156.

O'Neill, R.V., Hunsaker, C.T., Jones, K.B., Riitters, K.H.,Wickham, J.D., Schwartz, P.M.,Goodman, I.A., Jackson, B.L. and Baillargeon, W.S. 1997. Monitoring environmental quality at the landscape scale;

using landscape indicators to assess biotic diversity, watershed integrity and landscape stability. *Bioscience*, 47(8):513–519.

O'Neill, R.V., Hunsaker, C.T., Timmins, S.P., Jackson, B.L., Jones, K.B. Riiters, K.H. and Wickham, J.D. 1996. Scale problems in reporting landscape patern at the regional scale. *Landscape Ecology*, 11(3):169–180.

O'Neill, R.V., Johnson, A.R. and King, A.W. 1989. A hierarchical framework for the analysis of scale. *Landscape Ecology*, 3(3/4):193–205.

O'Neill, R.V., Krummel, J.R., Gardner, R.H., Sugihara, G., Jackson, B., DeAngelis, D.L., Milne, B.T., Turner, M.G., Zygmunt, B., Christensen, S.W., Dale, V.H., and Graham, R.L. 1988. Indices of landscape pattern. *Landscape Ecology*, 1(3):153–162.

Overton, W.S., White, D. and Stevens, D.L. Jr. 1990. Design Report for EMAP. EPA/600/3-91/053, U.S. Environmental Protection Agency, Office of Research and Development, Washington, DC.

Patil, G. P. 2002. Conditional entropy profiles for multiscale landscape fragmentation and environmental degradation. In: El-Shaarawi, A., and Piegorsch, W.W. (eds), *Encyclopedia of Environmetrics*, Volume 1, pp. 413–417. John Wiley & Sons, UK.

Patil, G.P. 2003. Overview: Landscape health assessment. In: Rapport, D.J., Lasley, W.L., Rolston, D.E., Neilsen, N.O., Qualset, C.O., and Damania, A.B. (eds), *Managing for Healthy Ecosystems*, pp. 559–565. Lewis Publishers, Boca Raton, FL.

Patil, G.P. 2004. Editorial: Special Institutional Thematic Issue: Center for Statistical Ecology and Environmental Statistics. *Environmental and Ecological Statistics*, 11(2):109-112.

Patil, G.P. 2005. Geoinformatic hotspot systems (GHS) for detection, prioritization, and early warning. *2005 Proceedings of the National Conference on Digital Government Research*, pp. 116-117.

Patil, G.P. 2006a. Digital Governance and Hotspot GeoInformatics for Monitoring, Etiology, Early Warning, and Management Around the World. *2006 Proceedings of the 7th Annual International Conference on Digital Government Research*, pp. 75-76.

Patil, G.P. 2006b. Special institutional thematic issue: Penn State cross-disciplinary classroom in statistical ecology and environmental statistics. *Environmental and Ecological Statistics*, 13(4):XX. In Press.

Patil, G.P., Acharya, R., Glasmeier, A., Myers, W.L., Phoha, S., and Rathbun, S. 2007. Hotspot Detection and Prioritization GeoInformatics for Digital Governance. In: *Digital Government: Advanced Research and Case Studies*. Chen, H., Brandt, L., Gregg, V., Traunmuller, R., Dawes, S., Hovy, E., Macintosh, A., and Larson, C.A., eds. Springer, Boston, MA. In Press.

Patil, G.P., Acharya, R., Modarres, R., Myers, W.L., and Rathbun, S.L. 2006. Hotspot geoinformatics for digital governance. In: *Encyclopedia of Digital Government*. Anttiroiko, A.-V. and Malkia, M., eds. Idea Group, Inc., Hershey, PA. In Press.

Patil, G.P., Balbus, J., Biging, G. JaJa, J., Myers, W.L., and Taillie, C. 2004. Multiscale advanced raster map analysis system: Definition, design, and development. *Environmental and Ecological Statistics*, 11(2):113-138.

Patil, G.P., Bishop, J.A., Myers, W.L., Taillie, C., Vraney, R., and Wardrop, D. 2004. Detection and delineation of critical areas using echelons and spatial scan statistics with synoptic cellular data. *Environmental and Ecological Statistics*, 11(2):139-164.

Patil, G.P., Brooks, R.P., Myers, W.L., and Taillie, C. 2003. Multiscale advanced raster map analysis system for measuring ecosystem health at landscape scale–A novel synergistic consortium initiative. In: Rapport, D.J., Lasley, W.L., Rolston, D.E., Neilsen, N.O., Qualset, C.O., and Damania, A.B. (eds), *Managing for Healthy Ecosystems*, pp. 567–576. Lewis Publishers, Boca Raton, FL.

Patil, G.P., Duczmal, L., Haran, M., and Patankar, P. 2006. On PULSE: The Progressive Upper Level Set Scan Statistic System for Geospatial and Spatiotemporal Hotspot Detection. Workshop, *The 7th Annual International Conference on Digital Government Research*, San Diego, CA.

Patil, G.P., Johnson, G.D., Taillie, C., and Myers, C. 2000. Multiscale statistical approach to critical-area analysis and modeling of watersheds and landscapes. In: Rao, C.R., and Szekely, G.J. (eds), *Statistics for the 21st Century: Methodologies for Applications of the Future*, pp. 293–310. Marcel Dekker, Inc., New York.

Patil, G.P., Modarres, R., Myers, W.L., and Patankar, P. 2006. Spatially constrained clustering and upper level set scan hotspot detection in surveillance geoinformatics. *Environmental and Ecological Statistics*, 13(4):XX. In Press.

Patil, G.P, Myers, W.L, Luo, Z., Johnson, G.D., and Taillie, C. 2000. Multiscale assessment of landscapes and watersheds with synoptic multivariate spatial data in environmental and ecological statistics. *Mathematical and Computer Modeling on Stochastic Models in Mathematical Biology*, 32:257–272.

Patil, G.P. and Taillie, C. 1979. An overview of diversity. In Grassle, J.F., Patil, G.P., Smith, W. and Taillie, C. (eds.). *Statistical Ecology, Vol. 6, Ecological Diversity in Theory and Practice*. International Cooperative Publishing House, Fairland, Maryland.

Patil, G.P. and Taillie, C. 1982. Diversity as a concept and its measurement. *Journal of the American Statistical Association*, 77(379):548–561.

Patil, G.P., and Taillie, C. 1999. Quantitative characterization of hierarchically scaled landscape patterns. In *Bulletin of the International Statistical Institute*, Volume 58, Book 1. pp. 89–92.

Patil, G.P. and Taillie, C. 2000. Multiscale frequency table analysis of landscape fragmentation in thematic raster maps. Technical Report 2000-0701, The Center for Statistical Ecology and Environmental Statistics, Department of Statistics, Penn State University, University Park, PA.

Patil, G.P., and Taillie, C. 2002. Multiscale frequency table analysis of landscape fragmentation in thematic raster maps. *Sankhya*, 64, Series A, Pt. 2:344–363.

Patil, G.P., and Taillie, C. 2004a. Upper level set scan statistic for detecting arbitrarily shaped hotspots. *Environmental and Ecological Statistics*, 11(2):183-197.

Patil, G.P., and Taillie, C. 2004b. Multiple indicators, partially ordered sets, and linear extensions: Multi-criterion ranking and prioritization. *Environmental and Ecological Statistics*, 11(2):199-228.

Pearson, S.M., Turner, M.G., Gardner, R.H. and O'Neill, R.V. 1996. An organism-based perspective of habitat fragmentation. In: Szaro, R.C. and Johnston, D.W. (eds.), *Biodiversity in Managed Landscapes, Theory and Practice*, pp. 77–95. Oxford University Press, New York.

Petersen, G.W., Hamlett, J.M., Baumer, G.M, Miller, D.A., Day, R.L. and Russo, J.M. 1991.*Evaluation of Agricultural Nonpoint Pollution Potential in Pennsylvania Using a Geographical Information System.* ER9105. Environmental Resources Research Institute, University Park PA.

Pielou, E. C. 1975. *Ecological Diversity.* Wiley-Interscience, New York.

Qi, Y. and Wu, J. 1996. Effects of changing spatial resolution on the results of landscape pattern analysis using spatial autocorrelation indices. *Landscape Ecology*, 11(1):39–49.

Rapport, D. J., Lasley, W. L., Rolston, D. E., Neilsen, N. O., Qualset, C. O., and Damania, A. B. 2003. *Managing for Healthy Ecosystems.* Lewis Publishers, Boca Raton, FL.

Rathbun, S. and Black B. 2006. Modeling and spatial prediction of pre-settlement patterns of forest distribution using witness tree data. *Environmental and Ecological Statistics*, 13(4). In Press.

Rathbun, S. and Fei, S. 2006. A spatial zero-inflated Poisson regression model for oak regeneration. *Environmental and Ecological Statistics*, 13(4). In Press.

Riitters, K.H., O'Neill, R.V., Hunsaker, C.T., Wickham, J.D., Yankee, D.H., Timmins, S.P., Jones, K.B. and Jackson, B.L. 1995. A factor analysis of landscape pattern and structure metrics. *Landscape Ecology*, 10, 23–29.

Riitters, K.H., O'Neill, R.V., Wickham, J.D. and Jones, K.B. 1996. A note on contagion indices for landscape analysis. *Landscape Ecology*, 11(4):197:202.

Ritchie, M.E. 1997. Populations in a landscape context: sources, sinks and metapopulations. in Bissonette, J.A. (ed), *Wildlife and Landscape Ecology, Effects of Pattern and Scale.* Springer, New York. pp. 160–184.

Roth, N.E., Allan, J.D. and Erickson, D.L. 1996. Landscape influences on stream biotic integrity assessed at multiple spatial scales. *Landscape Ecology*, 11(3):141–156.

Rodríguez, S. 2001. Statistical data mining of remote imagery for characterization, classification and comparison of landscape and watersheds of Pennsylvania. Ph.D. Thesis, The Pennsylvania State University.

Sekercioglu, C.H., Ehrlich, P.R., Daily, G.C., Aygen, D., Goehring, D. and Sandi, R.F. 2002. Disappearance of insectivorous birds from tropical forest fragments. *Proceedings of the National Academy of Sciences*, 99(1):263-267.

Stiteler, W.M. 1979. Multivariate statistics with applications in ecology. in Orloci, L., Rao, C.R. and Stiteler, W.M (eds.), *Multivariate Methods in Ecological Work.* International Co-operative Publishing House, Fairland, Maryland. pp. 279–300.

Townshend, J.R.G. and Justice, C.O. 1988. Selecting the spatial resolution of satellite sensors required for global monitoring of land transformations.*Int. J. Remote Sensing,* 9(2):187–236.

Urban, D.L., O'Neill, R.V. and Shugart, H.H. Jr. 1987. Landscape ecology; a hierarchical perspective can help scientists understand spatial patterns. *Bioscience,* 37(2):119–127.

Whitney, G.G. 1990. The history and status of the hemlock-hardwood forests of the Allegheny Plateau. *J. Ecology,* 78:443–458.

Wickham, J.D. and Riitters, K.H. 1995. Sensitivity of landscape metrics to pixel size. *International Journal of Remote Sensing,* 16(18):3585–3594.

Wiens, J.A. 1989. Spatial scaling in ecology. *Functional Ecology,* 3:385–397.

Wiens, J.A. 1995. Landscape mosaics and ecological theory. In: Hansson, L., Fahrig, L. and Merriam, G. (eds.) *Mosaic Landscapes and Ecological Processes,* pp. 1–26. Chapman and Hall, London.

Yahner, R.H. 1988. Changes in wildlife communities near edges. *Conservation Biology,* 2(4):333–339.

Yazvenko, S.B. and Rapport, D.J. 1996. A framework for assessing forest ecosystem health. *Ecosystem Health,* 2(1):40–51.

Index